T0120994

Acclaim for James Gleick's

Faster

"Nimble, smart, often funny, and—best of all—fast."
—*The New York Times Book Review*

"Fascinating and disturbing, amusing and informative, *Faster* is an eclectic stew combining history, academic research, and anecdotes drawn from the popular media." —*The Boston Globe*

"*Faster*'s short, jewel-like essays read like dispatches from a Xanadu of maximum efficiency." —*Newsday*

"Engaging." —*Los Angeles Times*

"Gleick offers his pointed analysis with refreshing irreverence."
—*Time*

"Gleick has done a magnificent job of outlining and defining the problem in a cogent and witty fashion; this book is an exemplar of thorough reporting." —*Chicago Sun-Times*

"Gleick has a great eye for today's transitions."—*The Village Voice*

"*Faster* is a wry, many-faceted meditation." —*Salon*

"Gleick's style is swift and slick, his chapters brief, and the book zips by like an Epcot monorail." —*New York*

"Trains a magnifying glass on our speed-driven world, illuminating the modern human's obsession with time and challenging a few myths. . . . Thank you, James Gleick."
—*San Francisco Chronicle*

James Gleick

Faster

James Gleick (www.around.com) is the author of *Genius: The Life and Science of Richard Feynman* (available from Vintage Books) and *Chaos: Making a New Science,* both of which were National Book Award nominees. He lives in New York.

Also by James Gleick

Chaos: Making a New Science

Genius: The Life and Science
of Richard Feynman

Faster

Faster

The Acceleration of Just About Everything

James Gleick

Vintage Books A Division of Random House, Inc. New York

FIRST VINTAGE BOOKS EDITION, SEPTEMBER 2000

 Published in the United States by Vintage Books, a division of Random House, Inc., New York, and simultaneously in Canada by Random House of Canada Limited, Toronto. Originally published in the United States in hardcover in slightly different form by Pantheon Books, a division of Random House, Inc., New York, in 1999.

Grateful acknowledgment is made to the following for permission to reprint previously published material: *Farrar, Straus & Giroux, Inc.* and *Faber and Faber Limited:* Excerpt from "A Girl in a Library" from *The Complete Poems* by Randall Jarrell. Copyright © 1969 and copyright renewed 1997 by Mary von S. Jarrell. Reprinted by permission of Farrar, Straus & Giroux, Inc. and Faber and Faber Limited, London. • *Random House, Inc.* and *Faber and Faber Limited:* Excerpt from "No Time" from *W. H. Auden: Collected Poems* by W. H. Auden, edited by Richard Mendelson. Copyright © 1941 by W. H. Auden. Rights for the British Commonwealth administered by Faber and Faber Limited, London.

The Library of Congress has cataloged the Pantheon edition as follows:
Gleick, James.
Faster: the acceleration of just about everything / James Gleick.
p. cm.
ISBN 0-679-40837-1
1. Time—Psychological aspects—Popular works. 2. Time—Social aspects—Popular works. I. Title.
QB209.G48 1999
529'.7—dc21 99-21640
CIP

Vintage ISBN: 978-0-679-77548-5

Author photograph © Beverly Hall
Book design by Johanna Roebas

www.vintagebooks.com

146122990

For Harry

IN MEMORY

NOT ENOUGH TIME

I'm going to kill myself. I should go to Paris and jump off the Eiffel Tower. I'll be dead. You know, in fact, if I get the Concorde, I could be dead three hours earlier, which would be perfect. Or wait a minute. It—with the time change, I could be alive for six hours in New York but dead three hours in Paris. I could get things done, and I could also be dead.

—WOODY ALLEN

Clocks cannot tell our time of day
For what event to pray,
Because we have no time, because
We have no time until
We know what time we fill,
Why time is other than time was.

—W. H. AUDEN

Contents

Contents

Faster

Pacemaker

You are in the Directorate of Time. Naturally you are running late. You hurry past a glass-paned vault in which the world's number-one clock is soundlessly assembling each second from nine billion parts. It looks more like a rack of computers than a clock. In its core, atoms of cesium vibrate with a goose-stepping pace so sure, so authoritative, so humbling—but your mind wanders. There is not a moment to lose. Striding onward, you reach the office of the director of the Directorate of Time. He is a craggy, white-haired man called Gernot M. R. Winkler. He glances across the desk and says, "We have to be fast."

The directorate, an agency of the United States military, has scattered dozens of atomic clocks across a calm, manicured hilltop near the Potomac River in Washington. Armed guards stand watch at a security gatehouse down below, mainly because the Vice President's residence occupies the same grounds. Once past their scrutiny you can walk alone up the long drive to the stately 150-year-old Naval Observatory, the first national observatory of the United States. Long ago a four-foot ball of Charles Goodyear's Gumelastic rubber hung from a mast atop the observatory dome and dropped daily at noon to signal the time. Now the signals come more quickly. The Master Clock consults with fifty others in separate climate-controlled vaults—cesium clocks and hydrogen masers powered by diesel generators and backup batteries. They check off the seconds as an ensemble and communicate continuously via fiber-optic cable with counterparts overseas. The clocks monitor one another, and individual devices can come on or off line as their performance warrants. Out-of-sync clocks reveal themselves quickly. Winkler offers an analogy: "It's like a court of law, where you have many slightly different stories and one wildly different story." When the plausible witnesses are chosen and assembled, their output is statistically merged, worldwide, at the Bureau International des Poids et Mesures, outside Paris. The American contribution is the largest.

The result is the exact time. The *exact* time—by definition, by worldwide consensus and decree. The timekeepers at the directorate like to quote the old saw (Winkler quotes it now): "A man with a watch knows what time it is. A man with two watches is never sure." Humanity is now a species with one watch, and this is it.

Through most of history, time was fixed by astronomical reference points—the Earth spins once, call it a day. No more. The absolute reference has shifted from the stars to the atomic beams in their vaults. Particles are steadier than planets. Never mind the uncertainty principle; it is the heavens that cannot be relied on. Stars drift. The Earth shivers ever so slightly. With the oceanic tides acting as brakes, the planet slows in its rotation by fractions of a second each year. These anomalies do matter, in a time-gripped age. To compensate, the official clocks must every so often perform a grudging two-step, adding an odd second—a "leap second"—to the world's calendar. Most often, leap seconds are inserted at the close of December 31. The New Year clicks in sneakily: 11:59:58 P.M., 11:59:59, 11:59:60 (!), 12:00:00 A.M., 12:00:01. The descendant of the Naval Observatory's old Gum-elastic rubber ball drops, studded with light bulbs, in Times Square. Elsewhere, astronomical observatories, television networks, and time-obsessed computer users make an adjustment to catch the leap second. Observatories have been known to get the sign wrong, ruining a night's sky-watching with the difference between +1 second and −1. As the Earth continues to slow, leap seconds will grow more common. Eventually we will need one every year, and then even more. Scientists could have avoided these awkward skips by choosing instead to adjust the duration of the second itself. Who would notice? That is what they did, in fact, until 1955. They defined the second as 1/86,400 of a real day, however long that was. The second had to lengthen a tiny bit each year. The atomic clocks were retuned as necessary. This did not trouble most of us, even subliminally, but it did start to annoy atomic physicists, because they needed a temporal measuring stick that would not stretch: *come on, a second is a second—give me a real* SECOND.

So here is the real second. Here the technologies of speed reach the ultimate. "Fifty years ago," Winkler says wistfully—he was a schoolboy in Austria—"we made measurements of a tenth of a second from day to day. That was great. Then more and more applications came in with greater refinements. It is like anywhere in life. When you have a capability, people find a use for that.

"Submarines have to surface for communications—they have atomic clocks," Winkler continues. "Television transmitters have atomic clocks. If you have two transmitters on the same channel, and you are between two cities, the picture will go up and down unless they are on exactly the same frequency. All good television stations have a rubidium clock." You are briefly aware of something incongruous about this exactitude—but the hyperprecision is all too familiar, all too closely in step with the rhythms of your more ordinary haunts.

We have reached the epoch of the nanosecond. This is the heyday of speed. "Speed is the form of ecstasy the technical revolution has bestowed on man," laments the Czech novelist Milan Kundera, suggesting by *ecstasy* a state of simultaneous freedom and imprisonment ("He is caught in a fragment of time cut off from both the past and the future; he is wrenched from the continuity of time; he is outside time . . ."). That is our condition, a culmination of millennia of evolution in human societies, technologies, and habits of mind.

The finicality of the modern timekeepers departs even further from our everyday experience—a fact cheerfully acknowledged here at the directorate. Particle physicists may freeze a second, open it up, and explore its dappled contents like surgeons pawing through an abdomen, but in real life, when events occur within thousandths of a second, our minds cannot distinguish past from future. What can we grasp in a nanosecond—a billionth of a second? "I tell you," Winkler says, "it wasn't on a human scale when we were measuring time to a millisecond, and now we are down to a fraction of a nanosecond." Within the millisecond, the bat presses against the ball; a bullet finds time to enter a skull and exit again; a rock plunges into a still pond, where the unexpected geometry of the splash pattern pops into existence. During a nanosecond, balls, bullets, and droplets are motionless.

Inhuman though these compressed time scales may be, many humans crave the precision. Internet users set their computers to update their clocks according to the directorate's time signal. The directorate fields millions of automatic queries each day. By pinging back and forth across the network, software called NanoSecond or RightTime or Clockwork or TimeSync or Timeset can correct for propagation delays along the phone lines between the atomic clocks and you. Free connections can be made to modems or to "time servers" with the whimsical pair of addresses, *tick.usno.navy.mil* and *tock.usno.navy.mil.* More crudely, anyone with a telephone can dial the Naval Observatory's Master Clock Voice Announcer, for fifty cents the first minute. The time-obsessed used to keep their watches accurate to within seconds; now they keep their computers accurate to within milliseconds.

Nanosecond precision matters for worldwide communications systems. It matters for navigation by Global Positioning System satellite signals: an error of a billionth of a second means an error of just about a foot, the distance light travels in that time. One nanosecond—one foot. That is a modern equivalence worth memorizing. Cellular phone networks and broadcasters' transmitters need fine timing to squeeze more and more channels of communication into precisely tuned bandwidth. The military, especially, finds ways to use superprecise timing. It is no accident that the Directorate of Time belongs to the Department of Defense. Knowing the exact time is an essential feature of delivering airborne explosives to exact locations—individual buildings, or parts of buildings—thus minimizing one of the department's standard euphemisms, collateral damage.

Few institutions are so intensely focused on so pure a goal. Keeping the right time brings together an assortment of technologies and sciences. The directorate's astronomers study the most distant quasars—admiring them for their apparent fixedness in the sky. A favored set of 462 quasars provides as rigid a frame as can be found. Meanwhile, the directorate has a team of earth scientists to study the slowing rotation and the occasional wobble—a problem that comes down to watching the weather, because the planet's spin varies each year with the wind blowing on mountains. In all, the scientists who control the clocks have achieved a surpassing precision. As the eighteenth century mastered the measurement of mass, and the nineteenth, with the establishment of international geodesy, conquered the measurement of distance, the even ghostlier quantity, time, had to wait for the technologies of the twentieth century.

The seconds pass here with a consistency that no pair of scales or rulers can match. The worst distortion that can accumulate, each day, remains proportionately smaller than a hairsbreadth in the distance from the Earth to the Sun—the equivalent of one second in a million years. "This is extremely important," Winkler says, the accent of his native Austria breaking through. His hand slashes through the air like an ax. "We want to be *exact*."

So synchronize your watches. Here are the pacemakers, the merchants of exactitude, the owners of the pulse in the global circulatory system. When the Lilliputians first saw Gulliver's watch, that "wonderful kind of engine . . . a globe, half silver and half of some transparent metal," they identified it immediately as the god he worshipped. After all, "he seldom did anything without consulting it: he called it his oracle, and said it pointed out the time for every action of his life." To Jonathan Swift in 1726 that was worth a bit of satire. Modernity was under way. We're all Gullivers now.

Or are we Yahoos?

Your eyes wander toward Winkler's wrist—what sort of watch would satisfy the director of the Directorate of Time?—but you cannot quite see it, as he asks: "Can you miss a plane by a millisecond? Of course not."

He pauses and adds with pride, "I missed one by five seconds once."

It has been noted by psychologists and airline managers alike that some people prefer to arrive at airports in plenty of time, keeping time to spare, so that they can have time on their hands in the lounge or kill time in the bar. Others cannot be happy unless they time their arrival so closely that, having dashed the last fifty yards to the gate, they race up the ramp, flash their boarding pass at the flight attendant, and slip into their seat with the thunk of the aircraft door fresh in their ears. Not a moment wasted. Perhaps these dashers, always flirting with lateness, are the victims of what some doctors and sociologists have named "hurry sickness." Then again, perhaps it is the seemingly calm, secretly obsessive early arrivers who suffer hurry sickness more.

Both types must be seeking peace of mind. One type can relax in the waiting lounge or even the check-in line, having minimized the risk of missing a flight. The other can hope to rest assured that they have minimized a different quantity: wasted time. Airport gates are not the only places where people like to flirt with lateness. But in their way they serve as focal points in the modern world, places where the technology and the psychology of hurriedness come together. Airport gates are where we contemplate the miraculous high speeds of air transport and the unmiraculous speeds associated with getting to air transport. One measure of twentieth-century time is the supersonic three and three-quarter hours it takes the Concorde to fly from New York to Paris, gate to gate. Other measures come with the waits on the expressways and the runways. *Gridlocked* and *tarmacked* are metonyms of our era: to be gridlocked or tarmacked is to be stuck in place, our fastest engines idling all around, as time passes and blood pressures rise.

We are in a rush. We are making haste. A compression of time characterizes the life of the century now closing. Airport gates are minor intensifiers of the lose-not-a-minute anguish of our age. There are other intensifiers—places and objects that signify impatience. Certain notorious intersections and tollbooths. Doctors' anterooms ("waiting" rooms). The DOOR CLOSE button in elevators, so often a placebo, with no function but to distract for a moment those riders to whom ten seconds seems an eternity. Speed-dial buttons on telephones: do you invest minutes in programming them and reap your reward in tenths of a second? Remote controls: their very existence, in the hands of a quickreflexed, multitasking, channel-flipping, fast-forwarding citizenry, has caused an acceleration in the pace of films and television commercials.

We have a word for *free* time: leisure. Leisure is time off the books, off the job, off the clock. If we *save* time, we commonly believe we are saving it for our leisure. We know that leisure is really a state of mind, but no dictionary can define it without reference to passing time. It is unrestricted time, unemployed time, unoccupied time. Or is it? Unoccupied time is vanishing. The leisure industries (an oxymoron maybe, but no contradiction) fill time, as groundwater fills a sinkhole. The very variety of experience attacks our leisure as it attempts to satiate us. We work for our amusement. *Five hundred channels* became a watchword of the nineties even before, strictly speaking, it became a reality. It denotes too much to choose from. And not just channels: coffees, magazines and on-line 'zines, mustards and olive oils, celebrity perfumes and celebrity rumors, fissioning musical styles and digitized recordings of more different performances of Beethoven's Fifth Symphony than Beethoven could have heard in his lifetime.

All humanity has not succumbed equally, of course. If you make haste, you probably make it in the technology-driven Western world, probably in the United States, probably in a large city—including, certainly, the most prosperous cities of Europe and Asia. Sociologists in several countries have found that increasing wealth and increasing education bring a sense of tension about time. We believe that we possess too little of it: that is a myth we now live by. What is true is that we are awash in things, in information, in news, in the old rubble and shiny new toys of our complex civilization, and—strange, perhaps—stuff means speed. The wave patterns of all these facts and choices flow and crash about us at a heightened frequency. We live in the buzz. We wish to live intensely, and we wonder about the consequences—whether, perhaps, we face the biological dilemma of the waterflea, whose heart beats faster as the temperature rises. This creature lives almost four months at 46 degrees Fahrenheit but less than one month at 82 degrees. "Technology has been a rapid heartbeat, compressing housework, travel, entertainment, squeezing more and more into the allotted span," notes the social historian Theodore Zeldin. "Nobody expected that it would create the feeling that life moves too fast."

It has created exactly this feeling. The time signal that flows from the Master Clock to its millions of clients drives a coordination of global activity impossible even a generation ago: mass communication and mass culture depend on it. The laziest among us have acquired a heightened awareness of time—by necessity. The modern economy lives and dies by precision in time's measurement and efficiency in its employment. If money is the visible currency of trade, time is its doppelgänger, a coin over which companies and consumers battle, consciously or unconsciously, with ever-greater urgency. You probably notice most assaults on your wallet. Do you notice when a business makes a grab for a few extra seconds of your time? You may contemplate your losses while you wait in the serpentine line at the airport ticket counter or navigate a six-minute telephone queue that has replaced a human who might last year have answered your question in a few seconds. In return, marketers and technologists anticipate your desires with fast ovens, quick playback, quick freezing, and fast credit. We bank the extra minutes that flow from these innovations, yet we feel impoverished and we cut back—on breakfast, on lunch, on sleep, on daydreams. Federal Express and McDonald's have created whole new segments of the economy by understanding, capitalizing on—and then in their own ways fostering—our haste. "Tired of working overtime?" ask scores of advertisements. A medication is marketed "for women who don't have time for a yeast infection"—as though slackers might have time for that. The defining quality of haste is only now coming into focus in our cultural mirrors, as in the *New Yorker* cartoons: (1996) man getting into cab—"And step on it.

This restaurant may be over any minute"; (1997) man speaking into telephone—"No, I *don't* have four seconds to talk." Even Bill Gates, with his abundance of money, his private jet, and his fast cars, complains, "It seems like the whole world operates in five-minute intervals." So who can escape their awareness of the pressure? Not David Letterman. "I'll try to be brief—we've done a lot of focus groups and people complain that I'm talking too much," he tells the *Late Show* audience. "They say it delays the show."

Pruning minutes and seconds and hundredths of seconds has become an obsession in all but a few segments of our society. In the spirit of Olympic swimmers shaving their chest hair, television networks are ever so delicately shaving the "blacks"—the punctuation marks between shows, when the screen fades momentarily to darkness. Brooke High School in the northern panhandle of West Virginia tries to save one minute per class break and thus welcomes its students to the age of speed. "Opening lockers, grabbing books, dodging people and racing across the school are all obstacles students face when trying to switch from class to class in only four minutes," one girl writes plaintively. "Not only do teachers issue tardies to late students, but some issue them if a student is walking in the door a second or two late, still breathing heavy from sprinting the halls."

Yet we have made our choices and are still making them. We humans have chosen speed and we thrive on it—more than we generally admit. Our ability to work fast and play fast gives us power. It thrills us. If we have learned the name of just one hormone, it is adrenaline. No wonder we call sudden exhilaration a *rush*. "Your life is lived with the kind of excitement that your forebears knew only in battle," observes the writer Mark Helprin. And: "They, unlike you, were the prisoner of mundane tasks. They wrote with pens, they did addition, they waited endlessly for things that come to you instantaneously, they had far less than you do, and they bowed to necessity, as you do not. You love the pace, the giddy, continual acceleration." Admit it—you do! Still, you have not truly explored the consequences of haste in our culture and in our daily lives. You hardly perceive the acceleration of art and entertainment: the changing pace of media from cinema to television commercials, which reflect and condition a changing pace in our psyches.

Instantaneity rules in the network and in our emotional lives: instant coffee, instant intimacy, instant replay, and instant gratification. Pollers use electronic devices during political speeches to measure opinions on the wing, before they have been fully formed. Like missiles spawning MIRVs, fast-food restaurants add express lanes. If we do not understand time, we become its victims.

"Time is a gentle deity," said Sophocles. Perhaps it was, for him. These days it cracks the whip.

Life as Type A

Can our bodies take the strain? We suffer anxiety. We suffer stress. And more. One of the young technocrat nerd-heroes of Douglas Coupland's 1995 novel, *Microserfs,* has a theory. "Type-A personalities have a whole subset of diseases that they, and only they, share," he explains, "and the transmission vector for these diseases is the DOOR CLOSE button on elevators that only gets pushed by impatient, Type-A people."

He can count on you to get the joke. Everyone knows about Type A. This magnificently bland coinage, put forward by a pair of California cardiologists in 1959, struck a collective nerve and entered the language. It is a token of our confusion: are we victims or perpetrators of the crime of haste? Are we living at high speed with athleticism and vigor, or are we stricken by hurry sickness?

The cardiologists, Meyer Friedman and Ray Rosenman, listed a set of personality traits which, they claimed, tend to go hand in hand with one another and also with heart disease. They described these traits rather unappealingly, as characteristics about and around the theme of impatience. Excessive competitiveness. Aggressiveness. "A harrying sense of time urgency." The Type A idea emerged in technical papers and then formed the basis of a popular book and made its way into dictionaries. The canonical Type A, as these doctors portrayed him, was "Paul":

> A very disproportionate amount of his emotional energy is consumed in struggling against the normal constraints of time. "How can I move faster, and do more and more things in less and less time?" is the question that never ceases to torment him.
>
> Paul hurries his thinking, his speech and his movements. He also strives to hurry the thinking, speech, and movements of those about him; they must communicate rapidly and relevantly if they wish to avoid creating impatience in him. Planes must arrive and depart precisely on time for Paul, cars ahead of him on the highway must maintain a speed he approves of, and there must never be a queue of persons standing between him and a bank clerk, a restaurant table, or the interior of a theater. In fact, he is infuriated whenever people talk slowly or circuitously, when planes are late, cars dawdle on the highway, and queues form.

Let's think . . . Do we know anyone like "Paul"?

This was the first clear declaration of *hurry sickness*—another coinage of Friedman's. It inspired new businesses: mind-body workshops; videotapes demonstrating deep breathing; anxiety-management retreats; seminars on and even institutes of stress medicine. "I drove all the way in the right-hand lane," a Pacific Gas and Electric Company executive said proudly one morning in 1987 to a group of self-confessed hurriers, led by Friedman himself, by then seventy-six years old. In the battle against the Type A jitters, patients tried anything and everything—the slow lane, yoga, meditation, visualization: "Direct your attention to your feet on the floor. . . . Be aware of the air going in your nostrils cool and going out warm. . . . Visualize a place you like to be. . . . Experience it and see the objects there, the forms and shadows. Take another deep breath and experience the sounds, the surf, the wind, leaves, a babbling brook." Some hospital television systems now feature a "relaxation channel," with hour after hour of surf, wind, leaves, and babbling brooks.

We believe in Type A—a triumph for a notion with no particular scientific validity. The Friedman-Rosenman claim has turned out to be both obvious and false. Clearly some heart ailments do result from, or at least go along with, stress (itself an ill-defined term), both chronic and acute. Behavior surely affects physiology, at least once in a while. Sudden dashes for the train, laptop computer in one hand and takeout coffee in the other, can accelerate heartbeats and raise blood pressure. That haste makes coronaries was already a kind of folk wisdom—that is, standard medical knowledge untainted by research. "Hurry has a clearly debilitating effect upon the tissues and may in time injure the heart," admonished Dr. Cecil Webb-Johnson in *Nerve Troubles,* an English monograph of the early 1900s. "The great men of the centuries past were never in a hurry," he added sanctimoniously, "and that is why the world will never forget them in a hurry." It might be natural—even appealing—to expect certain less-great people to receive their cardiovascular comeuppance. But in reality, three decades of attention from cardiologists and psychologists have failed to produce any carefully specified and measurable set of character traits that predict heart disease—or to demonstrate that people who change their Type A behavior will actually lower their risk of heart disease.

Indeed, the study that started it all—Friedman and Rosenman's "Association of Specific Overt Behavior Pattern with Blood and Cardiovascular Findings"—appears to have been a wildly flawed piece of research. It used a small sample—eighty-three people (all men) in what was then called "Group A." The selection process was neither random nor blind. White-collar male employees of large businesses were rounded up by acquaintances of Friedman and Rosenman on a subjective basis—they fit the type. The doctors further sorted the subjects by interviewing them personally and observing their appearance and behavior. Did a man gesture rapidly, clench his teeth, or exhibit a "general air of impatience"? If so, he was chosen. It seems never to have occurred to these experienced cardiologists that they might have been consciously or unconsciously selecting people whose physique indicated excess weight or other markers for incipient heart disease. The doctors' own data show that the final Group A drank more, smoked more, and weighed more than Group B. But the authors dismissed these factors, asserting, astonishingly, that there was no association between heart disease and cigarette smoking.

In the years since, researchers have never settled on a reliable method for identifying Type A people, though not for want of trying. Humans are not reliable witnesses to their own impatience. Researchers have employed questionnaires like the Jenkins Activity Survey, and they have used catalogues of grimaces and frowns—Ekman and Friesen's Facial Action Coding System, for example, or the Cook-Medley Hostility Inventory. In the end, nothing conclusive emerges. Some studies have found Type A people to have *lower* blood pressure. The sedentary and obese have cardiac difficulties of their own.

The notion of Type A has expanded, shifted, and flexed to suit the varying needs of different researchers. V. A. Price adds *hypervigilance* to the list of traits. Some doctors lose patience with the inconclusive results and shift their focus to anger and hostility—mere subsets of the original Type A grab-bag. Cynthia Perry finds that Type A people have fewer daydreams. How does she know? She asks them to monitor lines flashing across a computer screen for forty painfully boring minutes and finds that, when interrupted by a beep (1000 hertz at 53 decibels), they are less likely to press a black button to confess that irrelevant thoughts had strayed into their minds. Studies have labeled as Type A not only children (those with a tendency to interrupt and to play competitively at games) but even babies (those who cry more). Meanwhile, researchers interested in pets link the Type A personality to petlessness; a National Institutes of Health panel reports: "The description of a 'coronary-prone behavior pattern,' or Type A behavior, and its link to the probability of developing overt disease provided hope that, with careful training, individuals could exercise additional control over somatic illness by altering their lifestyle. . . . Relaxation, meditation, and stress management have become recognized therapies. . . . It therefore seems reasonable that pets, who provide faithful companionship to many people, also might promote greater psychosocial stability for their owners, and thus a measure of protection from heart disease." This is sweet, but it is not science.

Typically a Type A study will begin with researchers who assume that there are some correlations to be found, look for a wide variety of associations, fail to find some and succeed in finding others. For example, a few dozen preschool children are sorted according to their game-playing styles and tested for blood pressure. No correlation is found. Later, however, when performing a certain "memory game," the supposed Type A children rank somewhat higher in, specifically, systolic pressure. Interesting? The authors of various published papers evidently think so, but they are wrong, because if their technique is to keep looking until they find some correlation, somewhere, they are bound to succeed. Such results are meaningless.

The categorizations are too variable and the prophecies too self-fulfilling. It is never quite clear which traits *define* Type A and which are fellow travelers. The "free-floating, but well-rationalized form of hostility"? The "deep-seated insecurity"? "Their restlessness, their tense facial muscles, their tics, or their strident-staccato manner of speaking"? If you are hard-driving yet friendly, chafing yet self-assured—if you race for the airport gate and then settle *happily* into your seat—are you Type A or not? If you are driven to walk briskly, briskly, all the time, isn't that good for your heart?

Most forget that there is also supposed to be a Type B, defined not by the personality traits its members possess but by the traits they lack. Type B people are the shadowy opposites of Type A people. They are those who are not so very Type A. They do *not* wear out their fingers punching that elevator button. They do *not* allow a slow car in the fast lane to drive their hearts to fatal distraction; in fact, they are at the wheel of that slow car. Type B played no real part in that mass societal gasp of recognition in the 1970s. Type B-ness was just a foil. Doctors Friedman and Rosenman actually claimed to have had trouble finding eighty men in all San Francisco who were not under any time pressure. They finally came up with a few, they wrote solemnly, "in the municipal clerks' and the embalmers' unions."

Even more bizarrely, that first Friedman-Rosenman study also included a Group C, comprising forty-six unemployed blind men. Not much haste in Group C. "The primary reason men of Group C exhibited little ambition, drive, or desire to compete," the doctors wrote, "was the presence of total blindness for ten or more years and the lack of occupational deadlines because none was gainfully employed." No wonder they omitted Type C from the subsequent publicity.

If the Type A phenomenon made for poor medical research, it stands nonetheless as a triumph of social criticism. Some of us yield more willingly to impatience than others, but on the whole Type A is who we are—not just the coronary-prone among us, but all of us, as a society and as an age. No wonder the concept has proven too rich a cultural totem to be dismissed. We understand it. We know it when we see it. Type A people walk fast and eat fast. They finish your sentences for you. They feel guilty about relaxing. They try to do two or more things at once—read and watch television; shave and drive a car; climb StairMaster, watch television, *and* talk on cellular phone . . .

"Already 6:20, and books still uncatalogued, *Economist* unread," writes the essayist Cullen Murphy. " 'Slow down, you move too fast,/You got to make the morning last.' Words of song somehow percolate into sentience as I shave. Point conceded. Foot off the pedal. Deep breath, count to ten—0 . . . 1 . . . 0—urgently taking binary shortcut. Doesn't work, still revved up. Maybe Doc Friedman right after all."

Yes. No. And yet . . . Type A people really do press that futile button.

The Door Close Button

The elevator makes a suitable starting point because, among the many aggravators of Type A–ness in modern life, elevators stand out. By its very nature, elevatoring—short-range vertical transportation, as the industry calls it—is a pressure-driven business. Although there are still places on earth where people live full lives without ever seeing an elevator, the Otis Elevator Company estimates that its cars raise and lower the equivalent of the planet's whole population every nine days. This is a clientele that dislikes waiting. So consider the following object: the pressurized sky lobby.

The pressurized sky lobby is a room, high inside a mega-skyscraper, where passengers pass through an airlock to repressurize before making a rapid descent to the ground. The pressurized sky lobby does not yet exist, except in the dreams of elevator designers. There, it is a natural solution to a compelling problem. We arrive at it by the following route:

Mega-planners crave mega-skyscrapers—such developers as Donald Trump or, more plausibly in the 1990s, dozens of fast-growing, status-conscious Asian corporations with headquarters in the capitals of the Pacific Rim: Hong Kong, Tokyo, Kuala Lumpur, Shanghai, their dreams barely shortened by the region's economic collapse at the decade's end. By "mega" they mean two hundred or even five hundred stories.

The tightest constraint they encounter has nothing to do with steel or cranes or community boards or airport landing patterns. No—they are stymied by elevators. The bigger a skyscraper gets, the more volume must be set aside for banks and banks of elevators and their hoistways. Otherwise people will have to wait too long. One rule of thumb suggests a group of elevators for each fifteen floors. Another suggests that after sixty floors and four elevator groups, a transit point becomes necessary—a sky lobby. Some office towers have added double-deck cars, loading and unloading simultaneously on two floors, requiring from passengers an extra bit of intelligent concentration: are you heading for an odd-numbered or an even-numbered floor? If architects follow all these rules to the logical conclusion, they reach a result worthy of Kafka or Escher: a building that is all elevators.

So the scientists who specialize in short-range vertical transportation have designed smarter elevators—elevators with algorithms. They add microprocessors and program them with fuzzy logic—not just yes-no, stop-start. They install heat and weight sensors on cars and landings. There are elevators now that pack more computing power than a high-end automobile, which is to say more computing power than the Apollo spacecraft. In computer folklore, the elevator has attained a distinct status. In the early 1980s, computer people described it as "an archetypal dumb embedded-systems application" and "the canonical example of a really stupid, memory-limited computation environment."

Within a decade, as chip-based intelligence reached ever further down the hierarchy of inanimate objects, the mantle of Relatively Stupid Machine had passed to a more modest device, the toaster, and an elevator engineer could claim grandly, "We'll call Elevator Monitoring/Building Management phase II of the microprocessor revolution." With good programming, elevators have learned to skip floors when they are already full, to avoid bunching up, and to recognize human behavior patterns. They can anticipate the hordes who will gather on certain floors and start pounding the DOWN button at 4:55 P.M. each Friday.

But these refinements are not enough—not nearly—so elevators must go faster, too. Elisha Otis's original elevator traveled at eight inches per second. The fastest passenger elevators, mostly in Japan, travel at more than thirty feet per second. The record holder in the late 1990s was a special Mitsubishi elevator in a sightseeing tower in Yokohama: more than forty feet per second, a good climb rate for an airplane.

Then, elevator technology must honor not just the laws of physics but also the vagaries of human physiology. Advances in elevatoring have now bumped up against the limits of comfort; designers must worry about jerk rates and horizontal sway. There is also the raw force of acceleration: fighter pilots aside, people tend to be uncomfortable if elevators launch them or drop them at more than 1/8 the acceleration of gravity.

Improved materials and engineering have softened most of the bumps and grinds—Hitachi has developed a system of electromagnets to compensate for tiny side-to-side deviations in the vertical rails. But one small problem resists solution. Evolution neglected to armor the human eardrum against the sudden change in air pressure that comes with a fall of hundreds of feet at high speed. Natural selection rarely had the opportunity to work with survivors of this experience, to fine-tune their eustachian tubes in preparation for vertical transport. So at mid-century, when Frank Lloyd Wright designed a mile-high tower with 528 stories, helicopter landing pads, and quintuple-deck elevators running on atomic power, airline pilots instantly wrote to alert him to the impracticality. The age of high-altitude passenger aviation was just beginning, and the pilots knew that elevators descending thousands of feet within a minute or two would subject their passengers to severe inner-ear pain. Sure enough, decades later, the Sears Tower in Chicago had to slow its observation-deck elevators because at least one passenger had complained of a broken ear drum—an extreme manifestation of hurry sickness.

Thus a leading elevator consultant, James W. Fortune of Lerch, Bates and Associates, outlined an alternative solution: the pressurized sky lobby. "Passengers going to and from higher building floors and sky lobbies would transfer between the sky lobby by using interzone shuttles, getting a chance to depressurize and repressurize en route to their final destination," he wrote. "The advantage of this scheme is that the lift passengers could wait for the lifts in a prepressurization holding/waiting lock, and then board the lifts for a very rapid descent." While they wait, he added, they could fight boredom by watching audiovisual screens.

In reality Fortune cheerfully acknowledges the improbability of these office-tower compression chambers. Still, he says, "If we're really going to build these 150- and 200-story towers . . ." Or the 500-story Aeropolis 2001 tower proposed by Obayashi Corporation for Tokyo Bay. Not nearly so fantastic is the notion of the video screens meant to mollify, or stupefy, the waiting passenger. They are reminiscent of Dr. Friedman's Type A therapy sessions. "Fujitech has prototypes," Fortune says. "They're very attuned to calming music. They show you these scenes of cherry blossoms." Mirrors distract us, too, shrewd building designers have noticed.

Elevator companies, looking ahead, are planning to break their technological impasse by creating detachable cabs that will move horizontally as well as vertically. These will rise in one hoistway, slide off on tracks, and rise again through another hoistway, creating the effect of more elevators shafts. You will sit while you travel according to this scheme. Perhaps you will wear earplugs. You will relax.

Manufacturers need new technologies because the old technologies of short-range vertical transport seem to provoke humans to raw expressions of impatience. Anger at elevators rises within seconds, experience shows. A good waiting time is in the neighborhood of fifteen seconds. Sometime around forty seconds, people start to get visibly upset. "When they're waiting for an elevator, as well as when they're in an elevator, they don't really feel they can do much productive," says John Kendall, director of advanced technology at Otis. *Antsy* is the word Fortune uses (odd how we project our haste onto these steady-paced insects). Once on board, our antsiness only intensifies as we wait for the door to close. How long? *Door dwell,* as the engineers call it, tends to be set at two to four seconds. For some, that is a long time. And not just Americans. "If you travel in Asia at all, you will notice that the DOOR CLOSE button in elevators is the one with the paint worn off," says Kendall. "It gets used more than any other button in the elevator. When they're in the elevator they want to go." Japan has pioneered another feature, called "psychological waiting-time lanterns": as soon as someone presses a call button, a computer determines which car will reach the floor first and lights the appropriate signal well in advance of its arrival. This gives the illusion of an instantaneous response and, as a side benefit, herds riders into position for quick loading. They enter. Then, finally, as the door starts to close, the sight of a new passenger racing toward the elevator creates a moral test (stab the DOOR OPEN button, or feign obtuseness and look away?) which many riders fail to pass.

In the Directorate of Time, you hurried up the old foot-worn stairs. Studies by psychologists and sociologists have found a distorted time sense among people waiting for elevators, and distorted always the same way. If the subject says, "I had to wait ten minutes," the real duration might have been two minutes. Do elevators really cause us to abandon our basic ability to measure short intervals of time? Or do we choose to exaggerate for emotional effect? Two minutes' delay does not seem to justify comparisons with the torture chamber. If that delay didn't really last ten clock minutes, it reached ten minutes on some other scale. Researchers analyzing human behavior for Otis in 1979 watched closely as the seconds ticked by:

> Waiting, some stand still, others pace, and another may make small gestures of impatience such as foot tapping, jiggling change in a pocket, scanning the walls and ceiling with apparent concentration. . . . At intervals, nearly everyone regards the elevator location display above the doors by tipping their head slightly back and raising their eyes. . . . Men, but hardly ever women, may rock gently back and forth. . . .
>
> The long silences, the almost library hush, that we can observe where people wait for elevators are not only what they seem. . . . The longer the silence the more likely one or more of us will become slightly embarrassed, . . . the more embarrassing and tense are the little interior dramas that we play out each within our own theater of projection. . . .
>
> The actual period of waiting that elapses before a particular group may feel that waiting has become a nearly unendurable torment will probably vary significantly with the composition of the group, the time of day, and the type of building in which they are traveling. . . . The wait is hardly ever long, however much the subjective experience may stretch it out.

These researchers considered many possible explanations for the waiting elevator rider's anxiety. Our ground-roaming vertebrate ancestors, they note, had to learn vigilance against "airborne predators." Perhaps, standing in crowds of people peering edgily upward, we feel the vestigial, genetically encoded, neurochemical echoes of an ancient anxiety.

Then again, perhaps we're just in a rush.

The doors must close. Everywhere, transportation engineers are pressing to save tiny increments of time. Managers of New York City's subway system, not known for its clockwork precision, discovered that conductors were failing to enforce a rule that doors must close within forty-five seconds after they open. The effects cascaded through the system: a minute's delay for one train would cause backups half the length of Manhattan. To hurry passengers along, they tried installing signs that read "Step aside, speed your ride" and digital clocks relentlessly ticking off the allotted time. Then they tried ordering conductors to drop the word "please" from the phrase "Please stand clear of the closing doors." Similarly, the calculus of the elevator designers has doled out the precious seconds, taken them back, and doled them out again many times during the history of the past century. The first automatic rubber bumpers and electric eyes, meant to save us from being crushed by elevator doors, squandered time by letting the doors bounce open again. Engineers had to learn how much we could tolerate doors brushing against our clothing as we slipped by. The impact of a moving door increases with the square of its speed, sadly enough. Researchers concluded that human elevator operators were time-wasters in their own way—too polite. "Much time is lost by slow moving passengers who make no effort to hurry," said the president of Otis in 1953, in a pitch meant to sell his customers new, automated elevators. Those dratted passengers. "They know the attendant will wait for them. . . . But the impersonal operatorless elevator starts closing the door after permitting you a reasonable time to enter or leave." It was not just the elevators that would gain intelligence and efficiency. He added, "People soon learn to move promptly."And so we have.

Although elevators leave the factory with all their functions ready to work, the manufacturers realize that building managers often choose to disable DOOR CLOSE. Buildings fear trapped limbs and lawsuits. Thus they turn their resident populations into subjects in a Pavlovian experiment in negative feedback. The subjects hunger for something even purer than food: speed. A close cousin of DOOR CLOSE is the button attached to traffic signals at some crosswalks—what traffic engineers call the "push-to-walk" button. Most pedestrians suspect that this, too, has no function beyond allowing the impatient to identify themselves. At worst, this is true. At best, the push-to-walk button will eventually, within a minute or two, insert a "walk phase" into an otherwise walkless cycle.

How many times will you continue to press a button that does nothing? Do you press elevator call buttons that are already lighted—despite your suspicion that, once the button has been pressed, no amount of further attention will hasten the car's arrival? Your suspicion is accurate. The computers *could* instruct elevators to give preference to floors with many calls. But elevator engineers know better than to provide any greater incentive than already exists for repeated pressing of the button. They remember Pavlov. They know what happens to those dogs.

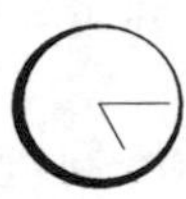

Your Other Face

While you wait, you look at your watch. It's a habit.

Where human anatomy meets data processing, there are just two important devices: the brain and the wristwatch. The brain is nice, but it doesn't tell time very well. So, creatures of habit that we are, we strap on that extra thing—a machine worn every day by nearly all adults in the industrialized world to display a single piece of information. "Your watch proclaims the essential you," says one radio advertisement of the 1990s. Your watch is "where you meet your other face." Anyway, we hold the time as closely as possible, where we can see it day and night. At night it glows.

Only after the spread of microchips did the mere telling of time turn out to be a limiting use of valuable anatomical real estate. Then technology began to outrun fashion. Inventors have produced wristwatches that announce appointments, or monitor pulse and blood pressure, or store phone numbers, or track the air or water temperature, or compute sums, or play music, *and* tell the time. There are emergency-beacon watches for pilots and switchblade-knife watches for James Bond wannabes. One designer created a prototype of a matchmaking watch: put your essentials in a database and the watch of a suitable romantic interest will blink when you draw nigh. Besides locating you in the fourth dimension, some watches can locate you, or at least orient you, in the first three—with altimeters, depth finders, and electronic compasses.

Watch technology lives by the ever-smaller, ever-faster technology of microprocessors. We have specialized computers in dishwashers and greeting cards; we may as well wear them on our wrists. A chip with the processing power of an early Apple PC fuels Timex's Data Link, a five-alarm appointment manager, to-do list, and phone book. Watch technology became a rocky shoal for pioneers of input-output ergonomics: the drawbacks of a tiny keyboard strapped to one hand are obvious, but they were not obvious enough for some manufacturers, it seems. Something, anyway, had to replace the ingenious classical ergonomics of wristwatches, the one-control interface technology that gave English the word *stem-winder.* Tiny batteries and motors replaced muscle-powered winding, and then buttons began to replace the stem and knob that had offered a fingertip-level feeling for the continuity and roundness of time. Buttons are cheaper and—as a watch gains more functions than buttons—paradoxically complex. Meanwhile, Timex's Data Link has its own approach to data transfer. It has an electric eye: point it at your computer and it grabs new appointments or phone numbers bit by bit from an eerie blinking bar on the screen. And the time, too—if you have made sure to use an Internet-based time-synchronizer to link your computer's clock to the Master at *tick-* and *tock.usno.navy.mil,* then the chain is complete.

Wrist technologists have suffered their share of anguish on the path to the perfect tiny device. Some watches have pressed the limits hard. Residents of some cities, for example, tested a Seiko "message watch" that made clever use of a sliver of FM radio spectrum to receive telephone pager messages, the closing Dow Jones Industrial Average, basketball scores, and the weather. AT&T itself announced a prototype of a Dick-Tracyesque phone watch and got the chairman of the Federal Communications Commission to pose with it for photographs. But the wrist phone's real-life prospects vanished, at least for the waning years of the twentieth century, because the marketplace failed to create a suitable cellular network. The predominant network uses antennas spaced as much as five miles apart, a design meant for relatively powerful car phones. Pocket phones can use the same network, but with short battery life. A wrist telephone would have much less broadcast power available and would therefore need more tightly spaced base stations—every few blocks in cities.

Nevertheless, the major watch companies in the United States and Japan have come to believe that smarter watches are the future. The exact time is a marvelous piece of knowledge to carry around on one's person, but not so marvelous as when the twentieth century was young. It has been cheapened by ready availability. Many men and women still treat a watch as jewelry and spend ten thousand dollars on the fine steel Swiss mechanical chronographs advertised in certain magazines, and more for less durable metals. The snob appeal is no longer in a watch's precision, as it was a generation ago, but rather in a nostalgic sense of craftsmanship. Any two-dollar quartz watch will keep better time. For fifty dollars you may feel entitled to more functionality. Snob appeal comes in many guises—now, perhaps, you want lunar phases, multiple time zones, and a date that will adjust itself correctly in leap years. Or perhaps you cannot feel truly complete until you wear a climber's wrist altimeter that stores fifty sets of daily climb data and shows a line chart of progress toward your target altitude, not to mention your maximum descent rate in ski mode. Or two centuries of tide predictions for 150 beaches worldwide. Or perhaps your kind of status symbol is the strapped-on minimini-computer that you have cleverly programmed to play Doom.

It remains easier to draw a new wrist technology in a comic strip than to sell it to consumers. More than a few flights of designer fancy have crashed against an invisible size-and-elegance barrier in the marketplace. Still, the new proliferation of wrist devices may be an early way station on a longer road. We strap these machines to our wrists because our bodies were not designed with many convenient niches for add-in devices, especially devices that we may need to see, hear, or speak into. But there are one or two other promising spots. For example, the bridge of the nose, with help from the ears, has been known to support a pair of optical lenses. Researchers at the Massachusetts Institute of Technology have developed a display screen sitting directly on a tiny chip—4 million pixels, finer resolution than most computer monitors, yet the size of a contact lens. Several companies are preparing commercial products with displays not much bigger: for example, a pocket-size fax receiver and viewer with a 1-square-inch screen.

The possibilities are frightening enough. Clever wristwatches might just be a way of beating around the bush—biding time till we implant these devices directly into our bodies. Once you've got the TV remote control conveniently strapped to your wrist, will you stop there? You're already way past pacemakers; you've seen the Bionic Woman and Robocop; you can stand pierced navels and nose rings. Maybe you're not too squeamish for what's next.

Whatever it is, it will tell us the time—the datum that rules our lives. In past centuries the bell that marked time synchronized the labor of sailors aboard ships and the contemplation of monks in abbeys. The industrial age required a more complex choreography of teams of workers with machines and so a more precise and authoritative timekeeping. As communication and transport reached out to unite cities and then countries, clocks provided the essential cohesion. For a ship at sea, carrying a clock made navigation possible. It was like carrying a precious flame under glass in a land without matches. The spread of clocks and watches linked people by getting their schedules into phase; thus timekeepers served as agents of social organization, enabling groups to live and work with the reliability of automata. Manufacture in bulk—by teams, as opposed to lone craftsmen—depended on regularity and repetition in process and supplies. Coordination spread by factory bell and whistle, then by company clock, then by individual watches. Karl Marx wrote Frederick Engels in 1863, "The clock is the first automatic machine applied to practical purposes; the whole theory of production of regular motion was developed through it." No wonder some historians describe the spread of timekeeping in terms of dehumanization and enslavement. "In the mechanized factory men are synchronized to machines, which in general have more regular habits than men," writes Sebastian de Grazia. "Materials too have to flow to feed the machines, and thus a synchronization of men, machines, and materials develops, more impersonal and complex than anything before. Most men today may not be aware that they are geared to machines—even while they are being awakened by the ringing of a bell and gulping down their coffee in a race with the clock."

But the emperors of ancient China used the measurement of time as a tool of authority. They reserved to themselves the right to establish a calendar and mark the passing hours with their elaborate, astronomical water clocks. They did not permit the spread of mechanical clock time that occurred a half-world to their west. "The Chinese treated time and knowledge of time as a confidential aspect of sovereignty, not to be shared with the people," notes the historian David S. Landes. The West turned clocks from imperial monuments into popular common property. And if our watches are slave-chains, we don them eagerly. Some may not care about high precision. Others, habitually late, set their watches ahead by five minutes as an exercise in self-deception (and can an integrated personality really trick itself into punctuality?). Still others care deeply about the finest of fine-tuning. When time-madness doesn't manifest itself as an overprecise watch, it might instead appear in the form of a multiplication of timepieces—time everywhere you look. "It's been 13 days, 8 hours, 23 minutes and 4 seconds since your last message according to the Compu-Mail electronic time," an E-mail correspondent writes in Avodah Offit's 1992 novel, *Virtual Love.* "There are three clocks on my desk and one on the wall. Lately I've taken to wearing an old pocket watch as a necklace pendant, in addition to my Swatch." How quickly convenience in getting the time leads to obsession in tracking it! When two of the time-obsessed meet, they bet on the accuracy of their watches and call 976-TIME. Europeans buy watches with internal antennas tuned to the frequency of a time signal transmitted from Germany. Similar watches are reaching the United States, and meanwhile hobbyists trade notes on How to Set Your Watch to Extremely High Accuracy.

"Never lose another 'whose watch is correct?' argument," advertises an applied mathematician. No one would apply the word *time-saving* to his synchronization procedure:

> You'll need a small screwdriver; if you have the jeweler's or eyeglass types, those are fine. When you open the watch, keep careful track of the various screws and plastic spacers. . . . Your target is something called a trimmer capacitor, a circular ceramic object that has a screwdriver slot in the top. . . .
>
> Keep a sheet of paper with times, notes, and sketches. . . .
>
> Wait a few days. Wear your watch as you always do, so the temperature changes are what the watch sees normally. Then estimate how fast or slow your watch is running. For example, suppose it's been three days elapsed time, or about 260,000 seconds. Your watch is 1.2 seconds fast, based on

> your eyeball guess. . . . So the watch's rate is 1.2/260000 or 4.6 parts per million fast. You'd like to get this under 1 ppm. . . .
>
> Try again; you may want to wait a week this time, to increase the accuracy of the rate-error estimate. . . .
>
> Keep doing this until things are to your satisfaction.

Somehow things never are. Still, to achieve one part per million precision in a twenty-dollar watch represents remarkable progress. For about the same twenty dollars, in the first part of the twentieth century, Britain's National Physical Laboratory, in Teddington, would test a watch, putting it through a forty-five-day trial, and award a Class A certificate if its pace varied less than five seconds a day, or about sixty parts per million. Nowadays the timekeepers at the Directorate of Time know from their leap-second experiences just how closely people pay attention. "You got a new watch at Christmas," Winkler says, "and you want to see how it does, and everything goes fine until January first." (Your eyes are drifting toward his wrist again. Did *he* get a new watch?) "Then we make this jump of one leap second, and people complain!"

For all its ubiquity the wristwatch is a new contrivance. People had tried attaching watches to, or hanging them from, armbands and bracelets, but wristwatches did not make any serious appearance until the late nineteenth century. Louis-François Cartier made one in 1904 to placate an aviator who had better things to do with his hands than draw out a pocket watch. Cartier attached a leather strap to one of his smaller models. Eventually it seemed that all humanity had its hands full in one way or another, or desired to know the time merely by glancing, or had just run out of little vest pockets, and within two generations the word *fob* had virtually disappeared from the English language. Watch chains became as outmoded as monocles. Into history vanished all the exquisite paraphernalia and body language of pocket watches. Some wearers regretted the brisk change. "The idiotic fashion of carrying one's clock on the most restless part of the body, exposed to the most extreme temperature variations, on a bracelet, will, one hopes, soon disappear," wrote a Professor H. Bock of Hamburg in 1917. The English upper classes, too, loved their pocket watches. When Sir H. Hardinge Cunynghame, former assistant undersecretary in the Home Office, wrote the definitive article, "Watches," for the *Encyclopaedia Britannica,* as late as 1936, he gave loving mention to their gilt and silver cases with death's-heads and other sacred emblems, to the enameled dials and musical boxes with performing figures, to the Skull Watch of Mary Queen of Scots and the book-shaped watch of Bogislaus XIV, Duke of Pomerania, and then, like a wonkish Martha Graham, he spelled out a choreography for when one's watch has accidentally been magnetized by proximity to an electric dynamo—"twirl it rapidly round while retreating from the dynamo and continue the motion till at a considerable distance"—and then he expressed concern about the growing use of artificial rubies and sapphires and the machine-stamping of metal parts, yet even then he could not bring himself to give so much as a mention to that upstart species, the wristwatch.

The wristwatch had already become more than a fashionable curiosity. It drove the miniaturization of machinery as the microchip would do again at the century's end. Fine-toothed gears shrank to a scale that tested the legendary eyes and lenses of Swiss artisans. The watchmakers squeezed jewels into holes a tenth of a millimeter or less, so perfectly machined that they required no adhesives. The first self-winding watch appeared in the 1920s, with a tiny weight swinging about the center of the movement to tighten the spring via reduction wheels. By mid-century, wristwatches were using the world's tiniest electric motors: a balance-hairspring oscillator driven by conventional electromagnets, still with mechanical contacts driving a gear one tooth at a time, or, beginning in 1953, the first true electronic engines, using tuning forks to jostle forward a feather-weight ratchet wheel.

Accutron was a milestone. This Bulova Watch Company tuning-fork watch, with a name that sounded jauntily modern in the 1960s, came into vogue as the latest in high-tech gimmickry and attracted a large share of the wristwatch market. The New York World's Fair selected an Accutron in 1964 as one of the touchstones of twentieth-century innovation and buried it in a time capsule, meant to be exhumed five millennia later. Astronauts wore Accutrons into space, inspiring new Astronaut Mark I and Astronaut Mark II models, to go with the Doctor's Date and Citizen Hi-Sonic. A philosopher of time, J. T. Fraser, wrote lyrically of the gamut of timekeeping processes: "pine cones in my study which open when ripe, geese migrating every spring and fall in response to some ancient call, the sun rising with great probability every morning, and the hum of my Bulova Accutron doing its regular 1.136,003,398,424 × 10^{19} cycles per sidereal year." At any rate, the curses of quaintness and irrelevance had now been cast upon the fine mechanical arts brought to their pinnacle of achievement in Switzerland. In 1968 the Swiss Council of State, facing humiliation from overseas, canceled forever the prestigious centuries-old competition in wristwatch chronometers. "The *régleurs*, those athletes of chronometry, bitter at their premature retirement, were left to nurse their memories," notes David Landes. "They meet now once a year for dinner; also at funerals, for this was a specialty that demanded experience, and these are men of a certain age."

No time to gloat, however. In 1970 obsolescence came to the Accutron itself, though it had 4,994 years still to go in the time capsule. Those fingernail-size tuning forks were silly, it turned out. They gave way to tiny slivers of quartz, carrying electric charge on their crystalline surfaces, driven to buzz at a constant frequency, usually a convenient power of two, such as 32,768 beats per second. Down the microscope hole again. Scientists had turned to the natural vibration of atoms. The Accutron's advertising boasted of its twelve moving parts, an amazing simplification of the traditional watch engine. A quartz watch with an LED display had no moving parts at all.

"Our quest for the precise time of day may go down in history as the greatest obsession of the twentieth century," notes the astronomer and anthropologist Anthony Aveni. True, for many of us the only useful function of the sweep second hand is to reassure us that our watches are running. When we confirm this, we see a formal indicator of the passage of time, at high speed—something the minute hand and hour hand could not show us. Next we look about for something to measure with this delightful tool. Among the most avid customers of the first watches with second hands were English horse-racing enthusiasts of the eighteenth century; by 1770, some dials went so far as to divide the seconds into fifths. Now we may time a pot on the stove, or check to see whether that television commercial lasts ten seconds or thirty, but most of our machines time themselves, from microwave ovens to cars.

Our watches serve community as much as chronology: the digits shining from the dial gain meaning only by reference to everyone else's watch. Call that enslavement to the routinized global economy if you prefer. The vestiges of the great Swiss watch industry survive because enough people still pay Audemar-Piguet and Patek Philippe and Vacheron & Constantin thousands of dollars for the most stylish possible machines. The market includes men who might not feel elegant wearing a brooch or a simple strand of pearls. Two centuries before, fashion suggested carrying watches in more than one pocket—some of them false watches, for the sake of economy. But watches are not just baubles. In them we see an icon of the ideal. The history of watch technology is a history of eliminating, one by one, each untidy imperfection, each possible source of disturbance, and each symptom of change. The watchmaker's challenge has always been to make a material object pure and platonic, immune from the vagaries of matter, the imperfections of the world.

"The clock is a machinery which repeats the same process over and over again," says Winkler at the Directorate of Time. "Now the same process means *undisturbed from the outside.* The observation itself is a disturbance, and we must keep that to a minimum. Magnetic fields, humidity. It's really technology driven to the utmost perfection in respect to control of a process." Like Felix Unger straightening up a room, artisans sought to eliminate the blemishes on their ideal of a machine. Inevitably, as they overcame the grossest sources of error, more subtle effects came into view. Metal expands when warm and contracts when cold—schoolchildren know that now, but in the seventeenth century Christiaan Huygens, creator of the pendulum clock, at first refused to believe reports that a pendulum would swing faster in winter. Clockmakers who measured their brass pendulums when cold and when warm found no difference in length—because they used rulers also made of brass. Generations later, they learned to compensate for thermal expansion with various devices, such as springs of two different metals bound together. Meanwhile they were bedeviled by friction. The art of watchmaking motivated an investigation into lubrication by fine oils from vegetable and animal sources. Watches came to require jewels, not as ornamentation but as wear-resistant machinery. Still new perturbations appeared: each tick of the drive train meant stress and vibration; a great technical achievement, the "deadbeat" escapement, dampened the recoils, and many other innovative escapements followed, until finally silicon let artisans abandon the vagaries of metal and oil altogether.

Eventually, if we pay attention to our watches, they teach us something even more valuable: that lived time is different from clock time. Our experience of time changes with our moods, with our age, with our level of busy-ness, with the complexity of our culture. We know how variable subjective time is, because we can so easily consult our icons of mechanical time—ornamental, utilitarian, objective time. So it is a version of the truth you wear on your wrist. "Although the ticks of the clock follow with trusted regularity as do the ax-time, sword-time, wind-time and wolf-time of the Vikings," says J. T. Fraser, "there is no way of knowing whether the next strike of Big Ben or the next oscillation of the quartz crystal will or will not occur." A lovely notion. But, poetry aside, modern watches are not so uncertain. You would be safe betting on the next oscillation of that quartz crystal, if you could—but it is already past before you can even form the thought.

Time Goes Standard

As long as a watch remains detached from the network—lacking a radio receiver for automatic updates—it is a bottle, like a thermos transporting volatile liquid oxygen, with thick walls keeping imperfection at bay. Its contents, the time, traveled as precious cargo on ships in the pre-wireless world. Without it, navigators could only guess at their longitude. To know the time in Greenwich, England, was to know one's place at sea.

The tendrils of fast communication spread at the end of the last century: telegraph lines, telephone wires into homes and businesses, and then the invisible broadcasts of radio signals. Almost from the first, networking meant synchronization. The single piece of information transmitted more than any other was the time—a "standard time." A few decades earlier, standard time had barely existed. Astronomers and ships' navigators concerned themselves with the time at the Royal Observatory, but the sprawling United States was a country of a thousand local times. Railroads changed that. Railroads demanded punctuality—they forced people to be "on the clocker" or even "on time." Until they could ride on trains, few people traveled fast enough to notice clocks set differently at their destination. It took telegraphs and telephones to synchronize clocks separated by hundreds of miles. In a networked world, time as a universal, ticking away everywhere in unison, seems normal, but to the nineteenth century, railroad time came as a shock—an unwelcome side effect of technology. It brought serious aftershocks—time zones, dividing neighbors along the boundaries, and daylight saving time, dividing city dwellers from farmers. Artificial, constructed, industrial-age time gave people a sense of its presumed opposite, *natural* time, a flow unbroken by machines, punctuated only by the swings or cycles of nature, and thus gentler in its effect on our true selves. Under the circumstances, even calendars started to look inhuman, with their unnatural partitioning of the flow of time. "The chopping up of time into rigid periods is an invasion of freedom, and makes no allowances for differences in temperament and feeling," wrote Charles Dudley Warner in *Harper's New Monthly Magazine* in 1884. Nor was he the first. Plautus had cursed the most advanced time-slicing technology he knew, the sundial: "The gods confound the man who first found out how to distinguish hours! Confound him, too, who in this place set up a sundial to cut and hack my days so wretchedly into small portions!" With the century ending, some towns and cities resisted the onslaught of precise and standardized railroad time. It was not until the end of World War I that the United States codified standard time in the law.

Synchronization across vast distances—standard time in the global village. The strangeness of it had not fully faded by the mid-twentieth century. *The Benny Goodman Story,* in 1955, captured it in a typical film gimmick: cutting from a radio studio, with four clocks for the United States time zones, to people dancing in each time zone—to the same music! Synchronization now brings a kind of entrainment of national biorhythms; residents of the American West Coast, for example, tend to rise earlier and go to bed later than East Coasters, because of the gentle pressure of communication across 3,000 miles. And synchronization also means acceleration, if only because the minutes and seconds are so tightly regulated. Henry David Thoreau saw this in the first days of railroad time—and he admired it. "Have not men improved somewhat in punctuality since the railroad was invented?" he asked. "Do they not talk and think somewhat faster in the depot than they did in the stage office? There is something electrifying in the atmosphere of the former place." Despite local unease, standard time became a coveted commodity. Town jewelers bought it from observatories by way of telegraphers for large annual fees and advertised it in their shop windows, gaining prestige. For short distances a time signal could be sent with low technology. Observatories dropped their balls, rang their bells, and blew their whistles. The spread of telegraph wires meant a spread in the authority of the national observatories in Greenwich, Paris, and Washington—disseminating a rhythm as steady as the drumbeat on a galley ship. Radio broadcasts took up the beat, and so, by the middle of the twentieth century, anyone with the right receiver could know the time to within a hundredth of a second. Then there were telephones. As people grew accustomed to having live communications devices in their homes, they naturally began to pick them up and ask for information. They called not just their friends but also an impersonal voice situated somewhere at the end of the line. It was as if the newly created telephone network had brought into the world a mysterious persona, all-knowing. In small towns people reached operators they knew by name, operators who sometimes listened in on their calls, but it was mostly city dwellers who demanded information from the impersonal telephone network. They asked for weather forecasts, election results, fire locations, and sports scores. Above all, they asked for the time of day. They did this in such numbers that the Bell Telephone managers, aware that they possessed no omniscient voice after all—just thousands of individual employees checking clocks and watches—put a stop to the practice in 1918. It cost too much in operator time.

This enraged customers. The time was not a luxury anymore; it was a necessity. "You have practically a monopoly," wrote David Elliot, a state senator from Colorado Springs—and a stockholder—to the telephone company president in New York. "You should return something for that monopoly outside the precise service the people pay for. Many more people are interested in the time of day than in the condition of the roads between here and Denver." He added, "P.S. I got the time of day from the Postal!" But telephone engineers, reviewing their records, estimated that in the years leading up to and just after World War I, as many as one call in forty nationwide was a request for the time. In big cities, the interest in the time of day was even greater. Time requests accounted for nearly 10 percent of the total telephone traffic in Chicago, for example. The whole enterprise inspired a certain self-consciousness about this desirable, salable commodity, the time of day. "All the way across the land," wrote a Rochester, New York, man, "the telephone of today has become the town clock of yesterday. . . . The householder who finds that his clock or watch has become erratic, or stopped entirely, is in a helpless position. Where but to the telephone may he turn for help in this emergency? . . . It is almost like a request for a drink of water."

It took a decade more—years of internal debate and assessment, calls counted by operators moving rubber pegs down the holes on their switchboards, memorandums sent carefully up and down the Bell chain of command—before the telephone companies finally decided that they could charge money for this service. Beginning in the summer of 1928 the New York Telephone Company and the New Jersey Bell Telephone Company created "Time

Bureaus," with a special number, Meridian 1212. The cost: five cents. On the first day alone, New Yorkers ponied up 10,246 nickels for this service. The Time Bureaus served as official ratification for that fictive, authoritative central voice, though it still amounted to nothing more than operators checking their clocks. When all this was still new, the *New York Times* philosophized editorially:

> Miserly souls may hesitate to spend a nickel for so trivial a purpose as an inquiry about the time, but we venture to predict that many will occasionally think the sum well spent. There are hundreds of heads of households whose duty it is to wind up the grandfather's clock every Sunday morning and see that it is set right. There are other thousands who forget to wind their watches before they go to bed, and must know when they wake up of a cold winter's morning how long they can stay in bed.

On the whole, however, city dwellers of the early twentieth century were not demanding the time just so they could go back to sleep.

The New Accelerators

Just knowing the time served as an accelerant. But there were others. Amphetamines—most famously methamphetamine—stimulate the nervous system, accelerate the heartbeat, and spark a fast-talking, restless feeling of excitement and energy. The inevitable slang name for such drugs: *speed.* Anything fast is said to be *on speed,* a metaphor within a metaphor. The linguistic variants have slid by so rapidly that you can be expected to understand when a car or a music video is said to be *meth-paced* or *methamphetaminic.* Athletes have used amphetamines as agents of literal speed (in vain); others have used them as dangerous antidotes to plain boredom. Meanwhile, narcotics offer a *rush.* And the last socially acceptable mood-altering drug in the puritanical 1990s, with alcohol and nicotine on the wane, turned out to be caffeine, the active ingredient not just in coffee but in new lines of soft drinks like Jolt and Surge. The Coca-Cola company promoted its Surge brand with the slogan "Feed the Rush." Pepsi fought back with newly caffeinated Josta and Pepsi Kona. Smaller companies began marketing mineral waters with caffeine and orange juice with caffeine; for example, Edge$_2$O and Edge$_2$OJ. Young men and women who once might have let a neighborhood bar soak up hours of their after-work time now settle for a few minutes at Starbucks. Connoisseurs on the run: that cup of joe might be Ethiopian Harrar-Moka or Sumatra Mandheling-Linton or the especially caffeinated Tanzania Peaberry. Just not Chock Full o' Nuts, thanks.

That was where Coca-Cola began, as a nineteenth-century tonic with caffeine as its secret ingredient. The company has maintained slyly that caffeine, though known to pharmacologists as an alkaloid that stimulates the central nervous system, serves in Coke merely as a flavor element. Never mind that it is a flavor element with no discernable taste. Coca-Cola was born in an era of tonics, restoratives, and bottled pick-me-uppers, and it was advertised as a cure for, among other things, "slowness of thought." Caffeine, we now know, can bring with it, in sufficient quantity, restlessness, nervousness, excitement, insomnia, flushed face, diuresis, gastrointestinal disturbance, muscle twitching, rambling flow of thought and speech, tachycardia or cardiac arrhythmia, periods of inexhaustibility, psychomotor agitation, and several other of the well-known conditions of our accelerated times. But don't worry—chew another chocolate-covered dark roast bean as you swing by the "Coffee à Go Go" Web site, which croons: "Caffeine, your friend and mine! Near and dear to our hearts, not to mention very tight with our synaptic impulses."

Even drinking alcohol and smoking tobacco have become speed-based pursuits. Liquor and cigarettes entered human life as time-savers, delivering their chemical effects far faster than wine and pipes had done. We *toss off* distilled spirits, notes the historian Wolfgang Schivelbusch, and thus we achieve more or less instantaneous intoxication, in contrast to the leisurely tippling of wine and beer. If we want to understand the progress of smoking technology from pipe to cigar to cigarette, he says, "what comes to mind is acceleration." We consume our stimulants faster and our depressants faster. It was thanks to cigarettes that the *smoke* became a unit of time, a quick five minutes or so, compared with the half-hour a cigar could consume— "as different," says Schivelbusch, "as the velocity of a mail coach is from that of an automobile." Cigarettes and shots of whiskey: quick little packages.

These are additives for our engines. We take them to modify the working of what we now quite consciously think of as the human machine. That's convenient, because it turns out that our knowledge of speed depends almost entirely on our knowledge of machines. Horses and falcons notwithstanding, it was only in the machine age that people became aware of speed as a quality that could be measured, computed, and adjusted. For the ancients, *speed* was indefinable. Before it meant velocity, Old English *spede* or *spēd* meant something more like success and prosperity; "God speed" didn't mean "May God hustle you along." Aristotle struggled enough with the abstraction of motion; to pin down a concept of velocity required a precision in measurement—and a belief in the precision of measurement—unattainable in the pre-Galilean, pre-Newtonian world. Languages had no words for the units of speed until the era of sail made necessary the quirky coinage *knot* (sailors measured their speed by heaving overboard a log tied to a rope and counting the evenly spaced knots as it played out). Even now, when the modern lexicon of units of measure includes joules and parsecs along with feet and pounds, the relative newness of speed shows up in a dearth of words: we almost always have to express speed in terms of a division of quantities: miles per hour, feet per second.

Before the machine age, few people had direct experience with uniform motion as expressed in Newton's equations. Steady speed first came with trains. The railroad bewildered passengers by causing familiar features of the landscape to float across their field of view *at high speed.* It did not take much speed to create amazing, strange sensations. "We flew on the wings of the wind at the varied speed of fifteen to twenty-five miles an hour," a first-time passenger wrote in 1830, "annihilating 'time and space.' " A mental leap was needed from speed as an attribute of planets and horses to speed as a variable, fine-tunable property. Machines let us make that leap. They gave us the everyday power to change a thing's speed by turning a dial or depressing a pedal. So why not change the speed of the human machine as well? If speed is what we want, why not take it by potion or tablet? In the nineteenth century more than a few pioneers in substance abuse (the euphemism of a later era) discovered that they could quite effectively alter the pace of the central nervous system, or at least create the sensation of altered pace. Arthur Conan Doyle's Sherlock Holmes, like quite a few real Londoners, relied on cocaine to shift from torpor to action. And H. G. Wells created a fantasy of drug-induced speed taken to an impossible extreme in a 1901 story, "The New Accelerator."

Wells's hero, a Mephistophelean-looking professor named Gibberne, seeks to discover a "nervous stimulant." Alice, in Wonderland, imbibed potions to change her size; Wells had a different dimension to explore. The times themselves were accelerating—"these pushful days" were his context—and ordinary people, "languid" people, needed a little something just to keep pace. So Professor Gibberne set out to discover an invigorator that would speed the nervous system by a factor of two or even three.

"Imagine yourself with a little phial," the professor says, holding up a green bottle, "and in this precious phial is the power to think twice as fast, move twice as quickly, do twice as much work in a given time as you could otherwise do." Power indeed, for a statesman under pressure of time ("he could dose his private secretary"), an author hurrying to finish a book, a doctor or lawyer with an urgent case, or a student cramming for an examination. And Wells adds one case that might not occur automatically to late-twentieth-century consumers of artificial stimulants: a duelist. "I suppose," the story's narrator muses, "in a duel—would it be fair?" The professor retorts, "That's a question for the seconds."

Wells, progenitor of modern science fiction, speaks for himself through his narrator: "I have always been given to paradoxes of space and time." His first literary triumph, *The Time Machine: An Invention,* laid out in fine philosophical detail the notion of time as a fourth dimension, no different in principle from the three spatial dimensions. This was five years before the turn of the century, and a decade before Einstein published his theory of special relativity. Space-time as a merged entity, a geometry, was already in the air. A little explosion of speculations about dimension came forth in England in the mid-1880s: books like Edwin Abbott's fanciful *Flatland,* which imagined a two-dimensional world and two-dimensional beings by way of inviting readers to add a fourth dimension to their world view. Albert Michelson and Edward Morley had conducted their famous experiment implying a constant velocity of light, and H. A. Lorentz had put forward his strange idea that a fast-moving object contracts in space. Wells surely followed these developments, but his prescient understanding of space-time geometry flowed as well from more mundane items in his modern world: new graphical representations of time in railway scheduling charts and weather histograms. These showed time as a fourth dimension, right there on the page. "Here is a popular scientific diagram, a weather record," says Wells's Time Traveler. "This line I trace with my finger shows the movement of the barometer. . . . Surely the mercury did not trace this line in any of the dimensions of Space generally recognized?" An efficient, measuring age made time recognizable as the fourth dimension before mathematicians and physicists worked out the details.

Time travel seems now like one of the most radical of science-fiction notions, in the sense that it remains unreachable as a technology, even after submarines, rockets, and ray guns have arrived on the scene. In an imaginative sense, however, the idea of traveling into the *future* was not so radical as all that. It is an act we perform at every instant. When we wake after a night's sleep, we wake into the future—into a changed world. Rip Van Winkle was a time traveler without a machine. Wells's *Time Machine* hero was a time traveler with ivory levers and quartz rods. And Professor Gibberne is a time traveler on drugs—"really preparing no less than the absolute acceleration of life."

He and the narrator mix the potion with water, clink glasses, and close their eyes. Moments later they sense that the world has slowed to a crawl. Long before the motion-picture industry made its own discovery of slo-mo, they open their eyes to a window curtain frozen in the act of flapping in the wind. Gibberne lets a glass slip from his hand, and they watch it drift ever so slowly toward the floor. Outside, motion has nearly ceased: a puff of dust hangs in the air behind a bicyclist, people leer in mid-gesture, and the narrator feels what a later time will call existential nausea at the exposed unpleasantness of it all—"strange, silent, self-conscious-looking dummies." Sounds from a shouting man and a band of musicians pulse slowly into their ears in the form of low rattles and sighs. They walk about at their thousand-times-accelerated pace, invisible to everyone else. They perceive the lazy motion of a horse's legs, a driver's whip, and a bumblebee's wings. They exult in the vantage point of their fast pace. Speed is power.

> To see all that multitude changed into a picture, smitten rigid, as it were, into the semblance of realistic wax, was impossibly wonderful. It was absurd, of course; but it filled me with an irrational, an exultant sense of superior advantage. Consider the wonder of it! All that I had said, and thought, and done since the stuff had begun to work in my veins had happened, so far as those people, so far as the world in general went, in the twinkling of an eye.

Then they begin to catch fire, their clothes scorching brown from friction with the air—Wells anticipating rockets in need of heat shields.

"The New Accelerator" reached forward into the new century to capture a set of feelings about speed: the force, the superiority, the efficiency, the sheer desirability. Gibberne's potion exaggerated only by degree the speededness that people were already seeking. It offered the ability to carve out pure slices of time for practical work. It promised hours to write or study, free from the pressing business of every day. It gave the taker an edge. It concentrated the utmost vigor. It bestowed invisibility as well as a new kind of sight. One could undertake anything, even criminal acts, "dodging, as it were, into the interstices of time." The narrator speculated about a counterpoint drug, a Retarder (opium? Quaaludes?) with a reverse effect, but true potency lay with the New Accelerator. We citizens of the next, meth-happy century live out a bit of New Acceleration not just when we ingest speed but also when we press our feet down on the accelerator pedal to slalom past the slower cars on any highway. We enjoy New Acceleration whenever, and however, we participate in what the novelist Nicholson Baker calls "time's cattle-drive."

Various modern television commercials play to the same time-stopping fantasy, showing, for example, at an airport, an executive retrieving her luggage and strolling to her rental car amid a crowd of momentarily frozen fellow travelers. Your Freudian analysis of this dream would reveal the wish of every business to speed past its competitors without suffering speed's harrying consequences. In reality, of course, we speed up, and they speed up, and competition continues, faster.

Baker himself wrote a variation on Wells's theme one hundred years later, a novel called *The Fermata.* Its hero can stop time just as Professor Gibberne could slow it to a crawl. Comically, he cannot imagine much to do with this power except undress women.

Where Wells thinks about the advantages for a statesman under pressure from the flow of events, Baker's narrator thinks about the usefulness for last-minute Christmas shopping. He performs his trick on the highway—"stopping the universe while driving at sixty miles an hour seemed an extremely rash and kinky thing to do"—and listens to his world not quite so carefully as Wells's narrator did: "I heard hooting and roaring noises in my ears when I walked into or away from the direction that I had been driving: I supposed it was something to do with vectors and frozen sound waves and the Doppler effect, but I didn't trouble myself over it." He doesn't trouble himself over much, beyond observing that sex itself is a way of stopping time. For him "the timelessness of the arrested instant" serves mostly as an escape, from tedious work or from fast-paced conversations he cannot quite understand.

"I've lost all conception of what 'soon' means," he tells a woman. "Don't you want to lose all conception of what 'soon' means, too?" We do, it seems.

Seeing in Slow Motion

The future in science fiction is usually just the present thinly disguised. The New Accelerator's inventor exclaims: "It kicks the theory of vision into a perfectly new shape!" As Wells was writing, the theory of vision was already being kicked into a perfectly new shape. A fresh way of seeing movement—by freezing time—directly inspired him. The sights that came to Professor Gibberne when he dragged the world around him into slow motion had already begun to come to a few pioneers. They used cameras.

Of the many omissions in the tapestry of human knowledge circa 1872, one especially peculiar gap was destined to bear on the next century's relationship with speed. No one actually knew whether a trotting horse lifted all four hooves from the ground at any point in its stride. Many equestrians thought they knew, but even those who were right were wrong. People had bred and raced horses for centuries, had painted them in realistic and fanciful poses, had watched them with large sums of money at stake, had created a nomenclature for classifying their distinct gaits—the walk, the trot, the amble, the rack, the canter, the gallop—but what were the horses actually doing? Their legs moved too fast for the human eye. Leland Stanford, former governor of California, founder of the university that bears his name, and an owner of race horses, wagered that a horse did fly free from the ground during its trot. Stanford had made a more prominent contribution to national synchronicity by presiding over the joining of the transcontinental railroad at Promontory, Utah. As an owner of horses, among them the famous Occident, he mainly wanted to win races, so he had more than an academic interest in the physiological details—whether the forelegs were straight when they touched the ground or whether the heel hit first. In 1873 Stanford engaged a transplanted Englishman, Eadweard Muybridge, to settle the matter for him.

Over the next five years Muybridge (interrupted by a brief murder trial—he had shot and killed his wife's lover) worked on a more and more complex apparatus beside the trotting track at Stanford's 700-acre estate in Palo Alto. Photography in that era meant huge wooden boxes on stands, with wet plates requiring long preparation and manually timed exposures. To stop the motion of Occident, who covered twenty feet in a single stride lasting barely a half-second, he tried a mechanical shutter, a moving board triggered by a spring as the horse passed. The first results were so shadowy that Muybridge had them retouched by a painter. By 1878, though, he had a line of twelve cameras with electromagnetic shutters, tripped by wires running across the track. The resulting series of pictures proved that a trotter did in fact lift all four feet from the ground. Soon after, another sequence dissected the gallop—even more startling. Painters had shown galloping horses arcing through the air with front and rear legs outstretched. In reality, a galloping horse does leave the ground, but not with its legs extended—rather, at the point in its gait when all four legs come together beneath its body. This was exciting. Muybridge's horse pictures appeared on the cover of *Scientific American* in strips meant to be cut out and viewed in a zoetrope, a new toy also known as a "wheel of life," a cylinder with slits around its circumference. When spun quickly, the zoetrope presented the images to the eye in sequence, creating the illusion of motion—recomposing Muybridge's fragments into a whole. People would pay to see the simplest gestures deconstructed and reconstructed, it seemed. Muybridge set to work studying the motions of athletes at the San Francisco Olympic Club and animals at the Philadelphia Zoo. He named his own projection device the "zoöpraxiscope" and took it on tour at fairs and expositions. "We predict that his instantaneous photographic, magic lantern will make the round of the civilized world," wrote the *Alta California* newspaper.

It did, and as one consequence it catalyzed painters. First, some critics sniffed. "We doubt whether the contribution to art will be of much importance," wrote the London *Globe.* "Art for the purpose of representation does not require to give to the eye more than the eye can see; and when Mr. Sturgess gives us a picture of a close finish for the Gold Cup, we do not want Mr. Muybridge to tell us that no horses ever strode in the fashion shown in the picture. It may indeed be fairly contended that the incorrect position (according to science) is the correct position (according to art)." Yet artists had always been willing to borrow from scientists to see more clearly. Anatomists' dissections had lent new accuracy to the muscular detail in nude portraiture, and now technology could cut open the movement, the gesture, and the action.

Muybridge had a counterpart in France: Jules-Étienne Marey, a physiologist at the Collège de France who had begun by inventing machinery to make graphs of the invisible rhythms of the heart. In the 1870s, as Muybridge was capturing the horse's gait, Marey invented similar cameras to break apart the equally invisible mechanics of the human walk. He was inspired to combine the successive slices of time on a single photographic plate, creating multiple images that came as revelations. With a "photographic gun," a camera employing a repeating shutter, he took analytical pictures of birds in flight. These two men, corresponding, quickly realized how many everyday actions had secrets to reveal, once technology let them see faster than could the naked eye. Their zoöpraxiscope and chronophotographie revealed a whole formerly unseen world: the serial grace of a woman carrying a bucket up steps or pouring water over her head, or letting her handkerchief flutter to earth; a shadow boxer; a nude blacksmith striking his anvil; a nude woman stepping up to a bed, drawing back the sheet, and climbing in—lifting her knee, bending sideways, sliding down under the covers, all this arrested in a few discrete frames; a child rising from the ground; a man changing a bayonet, pounding a mallet, opening an umbrella; acrobats, baseball players, high jumpers, fencers, lumbering elephants, dancers with lacy air-blown clothes or none at all. How little we knew! These days, when television cameras routinely display the spin of a four-seam fast ball, we take our temporal erudition for granted.

Photography thus began not just as a means of preserving a visual record or expressing an artistic vision. It froze a world in fast motion. It expanded the reach of our eyesight in the temporal domain, as microscopes and telescopes expand it in a spatial domain. Then artists, in turn, took on speed as a new mission. They tried to understand and reproduce fast motion as humans *seemed* to see it. "On account of the persistency of an image upon the retina, moving objects constantly multiply themselves," reported the 1912 Manifesto of the Futurist Painters. "Their form changes like rapid vibrations, in their mad career. Thus a running horse has not four legs but twenty." Famous paintings like Marcel

Duchamp's 1912 *Nude Descending a Staircase* tried to show the fragmented, multiple-image sensation that technology had exposed. But people don't really see motion that way, in blurs or strobe repetitions. Or they didn't, before photography. These paintings and photographs captured a hastening of perception. They managed to compress visual information in time as well as in space: impressionists, pointillists, cubists all playing at whittling the image to bare bones, finding its atoms and bits. They did this as ruthlessly as the painter Samuel Morse stripped language to dashes and dots. They learned to convey more with less.

They did not, however, capture the reality of speed—at least, not in any perfect or reliable way. The Futurists in their manifesto hoped for more: "The gesture which we would reproduce on canvas shall no longer be a fixed moment in universal dynamism. It shall simply be the dynamic sensation itself. Indeed, all things move, all things run, all things are rapidly changing." So they painted swallows in overlapping arcs, and bicycles in a near-abstract tumult of wheels. They loved cars. For that matter, they romanticized the machine gun. But taking a picture obliterates motion while revealing it. Soon, of course, the serial frames of Muybridge and Marey had their natural successor, the movie—the moving picture, the *motion* picture. With the movie, technologists had learned to reassemble speed. Until then, the slice of a gesture caught by the fast shutter was just one of a thousand fleeting states, not quite real and potentially ugly, like those revealed to H. G. Wells's experimenters in "The New Accelerator": the shards of sound they heard, the frozen leers they saw.

Another descendant of Muybridge and Marey advanced the technology of fast seeing much further: Harold Edgerton of the Massachusetts Institute of Technology. Edgerton began to experiment with flash tubes in the late 1920s, pulsing current through xenon gas and setting off bursts of light. So he stopped time, at least where the eye was concerned, and he, too, fell stroboscopically in love with the ordinary. Who could have guessed that a drop of milk, plunging into a saucer, breaks the surface tension, opens a crater, and sends two dozen new drops hurtling outward like jewels escaping the rim of a crown? Eventually Edgerton was called on to photograph the American atomic-bomb detonations. In the thirties his subjects were bullets passing through balloons and tennis balls rebounding from racquets. If Professor Gibberne had only known . . .

Now electronic devices can produce light bursts as short as thirty millionths of a second. The problem is to catch the *right* thirty millionths of a second. Experimenters use sound to pull the trigger. They can make fine adjustments to their timing by moving the microphone forward or back a few inches, taking advantage of the lethargic speed of sound through air. Light blasts from "flash bombs"; a high-velocity shock wave excites an instantaneous glow in gas for a millionth of a second or even less. The giant pulsed laser can produce light bursts as short as thirty nanoseconds. The extreme in ultrashort pulses has now reached down to the femtosecond range—a millionth of a nanosecond. These are the shortest events known to science. In a femtosecond, the Concorde flies less than the width of an atom. There is not much to see inside this temporal limbo, and only by clever inferences can scientists know just how short those pulses are. Then again, with the "femtosecond flashlight" available, condensed-matter physicists and chemists are thinking up applications—for example, getting an instant-by-instant understanding of what happens at the atomic level when a solid melts.

Meanwhile, every few years, in the flat deserts of Nevada or Utah, jet-powered automobiles strive to set a new record in a category increasingly irrelevant to our main interests, speed-driven though we are. The challenge for these cars is not so much attaining speed as staying on the ground without disintegrating. As the century opened, however, speed-lovers did see cars as the ultimate: "We declare that the splendor of the world has been enriched by a new beauty: the beauty of speed," announced the

Futurists. "A racing automobile with its bonnet adorned with great tubes like serpents with explosive breath . . . a roaring motor car which seems to run on machine-gun fire . . ." Filippo Marinetti wrote this in a hypermodern (and proto-Fascist) fever, and he had barely begun; a few years later he was talking about straightening out the Danube for the sake of high-speed river traffic.

But our speed is as much in how we see as how we move. When a racquetball hits a wall, there is a moment—less than a millisecond—when the leading and trailing edges of the sphere go inside out and meet at the center, so that the ball resembles a doughnut. A jet car travels a few inches. Scientists catch popcorn kernels in the act of bursting open. Waveforms appear in twanging rubber bands. Potato slices splatter chaotically against walls. Water drops fall into soap bubbles. Limes, bananas, and tomatoes explode as bullets drive through them. Hummingbirds freeze in flight.

"Nature refuses to rest," writes John Updike. He sees falling snow: "The transient sparkles seemed for a microsecond engraved upon the air." What we have learned to see, we can start to imagine. Only in an age of speed could we stop time. Only in an age of speed would we need to.

In Real Time

The fastest time is real time—a phrase we use casually now. It has a twisty, deceptively simple meaning. Real time doesn't just mean right now—hurry up, stat, on the bounce. Not long ago, all time was *real* time, but *real time* is no longer a redundancy. Some authorities call it a retronym, like *snail mail* and *acoustic guitar* and *rotary-dial telephone*: an old thing's new name, made necessary by the branching progress of innovation (E-mail and electric guitars and Touch Tone phones—and, what, artificial time? Imaginary time? Virtual time?).

But real time is not an old thing. Real time might be a limit we approach asymptotically, or it might be a state of mind. The *Oxford English Dictionary* tracks the origins of the phrase in the steamy primordial jungle of computerese:

1953 *Mathematical Tables & Other Aids to Computation* Vii. 73: With the advent of large-scale high-speed digital computers, there arises the question of their possible use in the solution of problems in 'real time,' i.e., in conjunction with instruments receiving and responding to stimuli from the external environment.

1960 *New York Times* 17 July 13/4: As an experiment, Air Force and Weather Bureau meteorologists attempted to use the pictures to make 'real time' forecasts of the weather—forecasts fresh enough to be useful.

1964 *Listener* 19 Nov. 784/1: [Computers] can more easily engage in activities in what we call 'real time.' That is to say, they can calculate at the actual speed of the events taking place.

1970 O. DOPPING *Computers & Data Processing* vi. 96: An example of a real-time process is a cheque account system in a bank where all transactions, e.g. withdrawals, are reported to the computer before they are finished.

As the OED's lexicographers discovered, *real time* began with the birth of computers. But computers did not create real time. Computers created fake time—simulated time in simulated realities. As they gained speed, the simulations began to catch up here and there with their real-world counterparts. The computer is defined by speed; it depends on speed, more than any of the fast machines that came before—more than the steam engine, more than the automobile, more than the airplane. Before nanosecond electronics, people could invent computers, as Charles Babbage and Byron's daughter Ada, countess of Lovelace, did in the nineteenth century, but the things were hardly worth making from tubes, let alone gears. Computers became practical only when semiconductors made them powerful enough (defined as *fast* enough) to create something that could be considered an alternative world—a mini-reality, a model of the world. This reality would have its own pace, of course. To be useful, a model of the Earth's weather would have to run just as fast as the original. Otherwise it would be forecasting yesterday's weather. A processing system for bank transactions had to run fast enough to handle real exchanges of money. That was when computer scientists needed a new phrase: *real time.* Real-time control of industrial processes means that the computer can actually keep up with the job, accepting input from sensors, carrying out calculations, and controlling valves or robots as the assembly line moves by. Real-time audio and video on the Internet means that the carrying capacity of your telephone line can keep up with the volume of data flowing from a live source—a presidential news conference or a planetary fly-by. Wonderful! Yet the phrase also carries a scent of irony these days, not always detectable on the printed page and not detected by the OED. Real-time oral communication is what used to be called *conversation.* Certain children of the nineties have been implored to put on their shoes "in real time."

There is real-time decision-making. "We're figuring it out right now, in real time, as we speak," an executive says, redundantly. The phrase is a now-ness intensifier. Real-time stock trading just means that your broker executes your order quickly, maybe so quickly that you don't have to wait for a call-back. Real-time scheduling, real-time cataloguing, real-time analysis, real-time auditing, real-time dance performance—all these mean a tiny bit more than just *fast* or *not too late.* Whatever real time is, we want it. Waiter, my entrée, please, in real time! In all of 1980 the *New York Times* used the phrase just four times. In 1990, thirty-one times—still protected, often, by the armor of quotation marks. By the end of the decade, *real time* appeared daily, up there with *cutting edge* and *quantum leap.*

Despite the bastardization, the concept of real time is a genuine addition to our understanding of haste. And not just haste. Real time implies communication. To understand any real-time process, we expand our sense of pace to include side-by-side time scales. With or without computers, we live complex lives. Maybe your finances are precarious enough to require tight synchronization with your creditors. If you can sign the check, lick the stamp, and toss the envelope onto the passing mail-room robot-cart just as you tell your caller, "It's in the mail," you have at least part of your life running in real time. The many embedded computers in an automobile work in real time. There would be no point in taking all that sensor input from wheels and brakes and calculating the likelihood of a skid that has already ended in a ditch. Real time means keeping up. A juggler performs computational brainwork in real time. A televised baseball game is broadcast in real time, which is to say "live," notwithstanding its many "instant" replays, which arrive in virtual time. The language of real time is a perpetual present tense: where a newspaper's sports columnist has the luxury of historical reflection, the quick-witted play-by-play announcer needs a syntax suited to real time, when present becomes past before our eyes, again and again, instant by instant. *Going . . . going . . . gone.*

The real-time transmission of data has been a twentieth-century obsession, fed by a parade of new technologies. Stock tickers were invented as dedicated machinery for the continuous timely display of a few crucial bytes; only later were their leavings parade-fodder. At first the essential live facts from baseball games were squeezed by the available bandwidth to the barest numbers, sent by telegraph to small-town radio announcers, who performed their own reconstitution of the data, filling in the dramatic play-by-play from their imaginations. Tensions over what could and could not be transmitted heightened as the decades passed. At century's end, privately owned sports enterprises filed lawsuits to assert control over the facts of their games—now being sent by ingenious entrepreneurs over the wires and airwaves to home computers and pocket pagers. The National Basketball Association sought to enjoin "the transmission of real-time data"—the running scores, leaking from the arenas, word-of-mouth made global and instantaneous. It was a futile effort. As Louis Menand commented, "This is a tough practice to enjoin, because human beings happen to be obsessed with real-time data."

The computer, of course, intensifies the obsession like a magnifying glass in sunlight. Then ten computers, connected, bring greater efficiency in the use of processing power. Then one hundred computers, connected, make campus or interoffice E-mail a live possibility. Then come new orders of magnitude, leading toward the modern Internet: roughly speaking, everybody's computers, connected. It is not just more; it is different. Chaos theorists understand such systems to undergo phase transitions, as water does when it turns coherently to ice or incoherently to steam. The controlling factor here is not heat or energy but pure connectivity. "We predict that large-scale artificial intelligence systems and cognitive models will undergo sudden phase transitions from disjointed parts into coherent structures as their topological connectivity increases beyond a critical value," wrote Bernardo Huberman and Tad Hogg at Xerox's Palo Alto Research Center in 1987, well before the birth of the World Wide Web. "At transition, these event horizons undergo explosive changes in size."

The crossing of one threshold turned the Web into a universal publishing medium. Suddenly—within five years of its creation—this crystallizing structure (or was it boiling?) threatened to subsume newspapers, bookstores, book publishers, radio and television stations, record producers, cartoonists, software distributors, and virtually every other purveyor of information. The slower media—annual journals, monthly magazines, daily newspapers, even radio and television—agonize over the disruption of their traditional cycles. The third round of presidential impeachment hearings in American history begins with testimony broadcast in real time—that is, live—but not on the major networks, which stay with their soap operas and talk shows in an attempt to salvage shrinking market shares. No matter. The public can choose in real time among many of the five hundred channels: CNN, *CNN Headline News,* two C-Spans, MSNBC, CNBC, *Fox News,* and CourtTV. During breaks, these stations will solicit the quickly forming opinions of viewers, to be delivered by telephone and Internet connections, and some of the instant commentary will note that the audience knew the testimony in advance, because it leaked out through myriad news channels during the night. Jeff Greenfield, for example, recalls nostalgically the days when nothing but quiet reflection disturbed the hours between the close of the evening news broadcast and the delivery of the next morning's newspaper. Now, he says, "The whole thing has been caught in this maelstrom of semi-informed, uninformed windbaggery." Peter Jennings began to catch up hours later on the formal broadcast of *ABC News,* which has been transformed by these upstart channels, as he said, into a more reflective elder statesman. Jennings himself was making the rounds of the five hundred channels to promote a book, a companion to a joint-venture television series on ABC and the History Channel. "CNN's going to be on all the time, and we're not going to be on all the time," he said. "And so, what we can do best, as CNN rushes through the day, is to try to give that which we do more meaning." He worries, now, about "the perils of going live"—the pressure that comes as technology removes the mandatory pauses:

> When I was a young reporter living in the Middle East I could go to India as I did, for example, in 1971 to cover the Indo-Pak War, and I would go off in the morning and cover something and then I would take the film and send it down

> to Dum-Dum Airport in Calcutta, and someone would hand-carry it to Bangkok if I were lucky, and then hand-carry it on to Tokyo where they'd process it and put it on the satellite to Los Angeles and land-line it to California—that's the process just to get the film back.

Then, of course, it seemed like rapid communication on the fast-shrinking globe. Now the same film is streaming out through Internet gossip sites before sunrise. In the twelve-hour gap Jennings could, so he believes, write his copy with more reflection and greater perspective than he can now—"today where I would just simply punch my camera in or my tape recorder in and just go live."

The information churns about, from one channel to the next, leaving whorls and eddies of self-reference. There are layers within layers, rapid reflection of analysis of punditry and quite a bit of round-the-clock regurgitation. Humanity may turn out to be a species, when digesting information, that chews its cud. Newspapers worry about being scooped by their own Web sites. In a bow to the medium, however, they might include faster and less-polished articles from the wire services, formerly treated as raw material for the newspaper's staff. Even these are too slow, as the newspapers also compete with nimbler 'zines and gossip services, feeding their unpolished thoughts into E-mail boxes. And some of these services, in snake-chasing-tail fashion, provide nothing more than digests of the newspapers themselves, but fast, and in quickly consumable pieces. Millions of amateurs rather than thousands of professionals perform these publishing functions. So suddenly all the traditional publishers scramble not to appear marginal and quaint. Then again, from the users' side, the faster experience has its flaws. There is friction implied by opening, folding, and turning the pages of a newspaper; or by choosing a book, cracking its spine, slitting its pages, adjusting the lamp, placing the bookmark; and this time-consuming frippery served an unintended purpose. Having made the investment, people found it natural to devote relatively large chunks of time to the actual reading. In contrast, the Web facilitates information consumption much as the remote control facilitated television watching. Reading on-line becomes another form of channel-flipping.

Nor is this skittish feeling confined to reading. The Internet similarly lubricates the trading of stocks. In the world before networking, a few stock-market obsessives—day traders—would congregate in the lobbies of brokerage houses, watching the ticker scroll by in electric lights. They would give buy and sell orders to a runner who would dash in and out. In the on-line world, day traders have come into their own. They have left the brokerage lobbies. They congregate by the hundreds or thousands in virtual spaces, where they shout their instant opinions by typing. To modern day traders, a stock bought and resold within seconds is short-term; a long-term transaction means someone has held a position for hours. And if by some chance a day trader hangs on to a stock overnight, it is probably a mistake and certainly a rule-breaker. Ken Wolff, a California day-trading instructor operating another Web site and chat room, tells his students that he hears too many excuses for holding stocks too long, like, "Heck, my stop was blown while I was in the bathroom." Overnight is even worse. Overnight, you're asleep!

The rise of electronic day trading flows from technologies that now allow virtually instantaneous execution of buy and sell orders on-line. For these quick exchanges of lightweight data, the brokerage commissions are cheap—often ten dollars or less. It used to cost enough and take long enough to buy a hundred shares of IBM that investors would actually have to think about it. Later they could bequeath them to their grandchildren. Not any more. Now, second by second, day traders are delighted if they can measure their profits in quarters, eighths, or "teenies." Here is what the Momentum Trader chat room, Wolff's site, sounds like on a given day:

> 7:22: Malichi just realized 500 shares of my PVTR 5/8 limit order got filled at 3/8. up an additional 1/4 on 500. :)
> 7:22: Cessna in XYBR 8 7/8 am I ok or bail?
> 7:23: thewoman sells PVTR +1/4
> 7:23: bbgun out ERGO at 6 3/16 late post
> 7:24: thewoman buys XYBR at 8 27/32
> 7:24: wwjd covers my short at 7 out flat and white as a sheet!

No one here cares about fundamentals. Day traders are not perusing balance sheets, analyzing business plans, or evaluating company management. They are staring at computer screens, watching numbers flash by, pressing their speed keys, and looking wishfully for "trends" and "patterns." They do watch for news, and their kind of news is shallow and brief—someone recommends a new drug, or quarterly earnings dip below analysts' expectations. A must for every day trader is a live, real-time, tick-by-tick quote system—formerly available only to professionals who paid well for the privilege, now available cheaply or even free all across the Internet. No one knows how many day traders there are—enough, anyway, to support a flood of overnight how-to books, teaching mills, and specialized brokerage firms. Many day traders seem to be students, if not exactly impoverished students, and many seem to be retired people, though sometimes the retirements are spur-of-the-moment; they have quit jobs that actually contributed to the world's economic life in favor of staying home and trading.

Their collective decision-making gives them a formidable mass to drive the prices of thinly held stocks. Their basic trading strategy is a weird mutation of the traditional "buy low, sell high"; day traders almost unanimously try to buy when a stock is rising and sell when a stock is falling. This strategy mirrors the current style of quite a few professional traders, and it is unstable by definition. The effect is to magnify stock movements, as day traders try to jump into hot issues and out of cool ones. Some stock swings—often where companies have some flavor of the Internet about their business—are startling, on monthly or even daily time scales. At the extreme, each rising stock becomes a pyramid scheme, where the last to buy in are the big losers. Market regulators began watching, at the end of the 1990s, for signs of growing volatility. They estimated that on-line trading already accounted for one-fourth of all transactions by individual investors. "It's hard enough to invest for the long term, let alone the nanosecond," said the Securities and Exchange Commission chairman, Arthur Levitt. It is hard enough to do anything for the long term on-line. The editors of *The Economist,* watching this new "cyber-army" from Great Britain, were even more skeptical: "The casino capitalists who spend seven or eight hours a day at their PC's trading Internet shares appear to be stark, staring mad." The regulators were slow to react to the new dynamics of a securities market driven by day traders—tiny lots of 100 or 500 shares, a peculiar focus on Internet stocks and fleeting bits of news, and astounding daily leaps in prices. In the day-trading rooms, hair-trigger connectivity turns momentary whims into mob obsessions.

The internetworking of the world's computers has opened more pathways through which all this data sloshes like a liquid of minimal viscosity. The data can take the form of software itself. This particular product of the industrial economy is made, conveniently, of bits. Because software can flow so quickly from a programmer's desktop to a consumer's, the computer industry has found its own timetables in a state of upheaval. David Hancock, chief of the Hitachi Corporation's portable computer division, drove his team with the slogan "Speed is God, and time is the devil." Software product cycles, already short at eighteen months to two years, have begun to evaporate. Instead of distinct, tested, shrink-wrapped versions of software, manufacturers distribute upgrades and patches that change within months or even days. They simultaneously offer official versions and more advanced versions still undergoing testing. With the Internet removing delays from the promotion and distribution processes, software manufacturers began to think in terms of six-month product cycles. Internet lifetimes were supposedly measured in "dog years." Kathleen Eisenhardt, a Stanford professor, advocated for business a sense of time like basketball's: "a fast, fluid game." That means no more five-year plans. "Five-year plans???" she said. "For managers who get it, it's more like a five-week schedule inside a five-month plan inside of a fifteen-month intuition."

Computer makers are reaching the extreme edge of what management consultants now call "fast cycle time": the systematic shortening of each step from conceiving a new product to delivering it. And Silicon Valley venture-capital firms have begun to seek fantastically short life-cycles for the companies they finance: eighteen months, they hope, from birth to public stock offering. Competition in cycle times has transformed segment after segment of the economy. In the automobile industry, development cycles traditionally spanned five years, until Japanese companies assembled a package of techniques to compete faster—including "just-in-time delivery" of parts from subcontractors. By 1993 the cycle was down to thirty-nine months; by 1997, twenty-four months, but Toyota was boasting eighteen months and trying for fourteen. The acceleration was not so hard—just a matter of removing layers of superfluous managers and bringing computers into the design process. Anyway, Detroit had no choice: customer tastes were changing faster and faster. Computers facilitated these speedy techniques and depended on them, too. Dell claimed in 1999 that parts spent an average of just eight hours in its factories before leaving as finished PC's. Now businesses are experimenting with "just-in-time accounting" (one step ahead of the auditors) and "just-in-time training" (only short-term memory required, presumably).

Is that bad? It can be a source of stress for executives who remember the old leisurely ways. Still, just-in-time inventory means less capital wasted in an inefficient supply chain and less risk of missing targets created by inaccurate, out-of-date forecasts. In the evolutionary competition for business fitness, the fast has driven out the slow. Sometimes the consumer benefits. Sometimes everyone just scurries around in real time. As recently as the 1970s, family snapshots required a week or so to develop, with prepaid Kodak mailers; then Fotomat stores began to appear, taking advantage of cheaper, smaller film-processing machinery to promise one-hour service. Names of the Fotomat descendants leave no doubt about the principal message: in Britain, say, we find Kallkwik, Prontaprint, and Snappy Snaps. Domino's pizza (half-hour delivery), Citibank mortgage lending (fifteen-minute approval), and countless other products and services have made speed the essence of their business strategy. Frito-Lay's sales force of ten thousand fires back daily reports through hand-held computers to the company headquarters. A study widely quoted among management consultants found that products coming in 50 percent over budget are far more profitable than products coming in six months late. Yet, a generation ago, it was far from obvious that speed pays. When Federal Express entered the marketplace, the older parcel-delivery services were baffled by the high prices it proposed to charge. The traditional pricing models were based on just two variables: weight and size. Speed, anyone?

Lost in Time

Even before the Internet, the hardware side of the computer business found itself racing forward under the whip of one of the century's essential laws of speed: Moore's law. The semiconductor pioneer Gordon Moore predicted in 1965 that chip density—and thus all kinds of computing power—would double every eighteen months or so. This has been correct. Moore's law codified our lightning speed-up in the pace of technological change. The acceleration of technology became exponential, officially. Seen on a scale of centuries, the speed-up began long ago. The printing press swept through the modern (postmedieval) world—over many human lifetimes. No wonder, later, that our great-grandparents thought telephones spread quickly! They and their children saw the changes wrought in society, instantaneously, it seemed, by the arrival of radio, then television. Now we see how gradual was the diffusion of those advances. Someone who first heard the geeky-sounding word *Internet* in, perhaps, 1994, and by 1995 was coolly surfing the Web night after night, could be forgiven for losing sight of the real speed of changing technology—how fast it is now, how slow it is still.

Let's say you are watching television, on a flat-screen active-matrix LCD screen that just hit the marketplace, or just on an "old" color set, fed its five hundred channels by a digital satellite service that sprang into being and signed up its first million customers within months, and either way the set is attached to a videotape recorder, not Betamax, of course, because that format was obsolete almost as soon as it was born, but Dolby-capable, and not just old Dolby but Dolby Surround, Dolby Pro-Logic, in fact Dolby Digital AC-3, ready for DVD and HDTV, bypassing altogether the dead end of the laserdisk player—anyway, you are watching TV, and you are not fiddling with rabbit ears, because you barely remember what rabbit ears were, and, whoa, this is black and white! Children tend to disdain any show in black and white, because they suspect it will be *too slow.* But this show is strangely familiar. The opening titles roll, a spaceship is about to take off for planets unknown, the mission control room is teeming with people, and the date appears on the screen in bold type. It is a year far, far in the future: 1997.

"This is the beginning," intones the narrator. "This is the day. You are watching the unfolding of one of history's great adventures." It's the unfolding of the first episode of the television series *Lost in Space,* as it aired on CBS in 1965, in the salad days of the American space program. One of your channels, the Sci-Fi Channel, is rerunning all of *Lost in Space* in it its original black-and-white splendor.

So there, on screen, photographers are milling about in Alpha Control to capture this historic 1997 moment, and flash bulbs pop. Yes, flash *bulbs.* Remember them? Actual bulbs that fired once, smelled of burnt metal, and had to be replaced for the next shot? Now the view pans back from an image of the galaxy displayed on a giant mockup of a vintage 1965 television set—the round corners are the giveaway. The control-room walls are lined with vintage 1965 computers: refrigerator-sized cabinets with flashing lights and large reels of tape. Sitting on one desk is an overhead projector, the kind last seen in your high school's audio-visual closet. It seems absurdly bulky. In fact everything looks bulky and somehow slow—the big old-fashioned television cameras rolling about with their cables trailing behind; even the lights (no miniature LED's). The mission engineers work at desks that seem to be equipped with big switches and bulky knobs and flashing lights of their own—but, you can't help but notice, no keyboards, no mice, and no display screens.

And what are those shiny round disks resting on the desks of this advanced, high-tech, space-mission control room? Ashtrays.

It's a tough job, predicting the evolution of technology, but somebody's got to do it. The production designers and art directors of science fiction over the past generation have had to go out onto one limb after another—and unfortunately the future does tend to arrive, sooner than expected. Who can guess what new materials, what new fashions, and what new devices will betray this year's science fiction when another generation has passed? Or what new social behavior: no one guessed, a generation ago, that a 1997 control room would be a no-smoking zone.

Norman Garwood made a clever end run around this problem in designing Terry Gilliam's 1985 film, *Brazil.* He created a glittery, sinister future filled with ancient technology—pneumatic tubes, teletype machines, desk spikes. The effect was a dark hodgepodge of the antique and the futuristic—perfect, because when the future does come creeping in, this is how it looks. It is not shiny and gleaming, neatly assembled in clean shrink-wrap. It comes all mixed up like a junkyard, the old and the new jumbled together. Your cellular phone and your portable CD player (its miniature laser not even worth a passing thought) rest on the shelf next to a 1950s cellophane-tape dispenser that somehow never got replaced. Or maybe you have one of those ancient, black, rotary-dial telephones, with handsets as heavy as stone, available now for premium prices at antique stores. There's one in the *Lost in Space* control room, and we aren't supposed to think "antique!" But how can we help it? When it rings, it *rings,* with the quaint old sound of a metal hammer striking a metal bell. "For its time, the hardware looked pretty sophisticated," recalls the actress June Lockhart, who hopped from planet to planet as Maureen Robinson during the series' three-year run. "Other than rotary phones and things like that." So step aboard the *Jupiter 2*: "the culmination of nearly forty years of intensive research and the most sophisticated piece of hardware yet devised by the mind of man," the narrator intones. Don't look too closely at the hardware-store fire extinguisher hanging on the spaceship wall; in a little while someone will be trying to throw it through the windshield. Don't think about the onboard clock, with its digits hand-painted on wheels that turn like a car odometer's. When an astronaut needs to step outside for a space walk, his tether turns out to be a rope made of the same fraying hemp that a sailor would have expected three hundred years ago. An up-to-date mountaineer can smirk.

For that matter, Lockhart also has nostalgic memories of the plastic laundry basket she used to carry around on the show. It was 1965 standard issue. "Then I would take the laundry and put it in a device and push some buttons and in fifteen seconds all the laundry has been washed, cleaned, and wrapped in plastic," she says. "So that is an innovation the world has yet to enjoy." In a few other ways our technology lags far behind the 1965 imaginations of the creators of *Lost in Space.* They could see a real-life moon landing just around the corner. Surely it was not too ambitious to guess that three decades later we would be able to launch a half-dozen people plus a goofy-looking robot in a two-story spacecraft.

But no. While computerization, miniaturization, and the technologies of electronics and artificial materials have all leapt forward, we find ourselves more stolidly earthbound than most scientists would have predicted. The technologies changing so rapidly have been those subject to Moore's law. They have brought not just the gross changes of a computer culture but also the subtler changes of texture that come with the pervasive miniaturization of lights and buzzers and miscellaneous conveniences in the fabric of our lives. It was easy to imagine interstellar space travel in 1965; it was more difficult to imagine that those future beings, toward the end of the century, would have telephone answering machines, fax, E-mail, and computers the size of credit cards. Even within the computer culture, people tend to overstate the importance of advances in heavy computing: the next leap forward in the megaflop, gigaflop, teraflop calculating performance of a supercomputer will not transform your life; the inexorable advances in small-computer connectivity will.

By comparison, the technologies of travel through our universe have barely changed. As of 1972, at the end of the American exploration of the moon, a total of twenty-four humans had made it farther from the Earth than the low, skimming orbit of, for example, the "Space Shuttle." As the millennium ends, the total is still frozen at twenty-four.

And the effect of all that rapid technological change? We get dizzy. We feel the instability of our own place in society. Soon, says the science journalist James Burke,

> the rate of change will be so high that for humans to be qualified in a single discipline—defining what they are and what they do throughout their life—will be as outdated as quill and parchment. Knowledge will be changing too fast for that. We will need to reskill ourselves constantly every decade just to keep a job.

We mistrust our machines all the more because we fall behind in learning to use them. It was amusing when fast-changing fashion meant rising or falling hemlines in Paris and Milan; it is stomach-churning when it means the wholesale replacement of LP's by CD's and then DVD's. We don't have time to read the manuals. Manufacturers, in their rush to market, don't have time for some of the niceties of consumer-friendly design, so the flashing 12:00 on the VCR becomes a cliché.

We don't run away from our progress-sustaining, life-enriching new technologies, but we understand when Edward Tenner, author of *Why Things Bite Back: Technology and the Revenge of Unintended Consequences,* pleads for a "retreat from intensity." Every software upgrade may not be worthwhile; higher workloads may not lead to more profit. Try "finesse," Tenner suggests. "In the office, finesse means producing more by taking more frequent breaks," he explains. "On the road, finesse means a calmer approach to driving, improving the speed and economy of all drivers by slowing them down at times when impulse would prompt accelerating."

On Internet Time

So Dean Hughson, a forty-seven-year-old egg merchant based in Las Vegas, Nevada, is sending E-mail to a colleague who he thinks is in Amsterdam, though he doesn't really care, and on the telephone another colleague is calling from Mexico City to say that he's about to send a fax via a new Internet service, and just at that moment his Seiko MessageWatch beeps because someone is paging him. Does it even matter that his wife is trying to get his attention, and she's in the same room? Then the cat enters the fray, jumping onto his desk, and Hughson looks at his life in wonderment: "Where in the hell is this old egg guy, and why is his life getting technologically more difficult as he gets older instead of easier?" Yet the forces of evil are not exactly dragging him willy-nilly into the information-flooded future. Hughson is, after all, the selfsame Webmaster who placed a specially designed button on his Internet site to allow anyone at all to flash him messages via his wristwatch. He waxes nostalgic about the slow-paced good old days, but something makes you wonder how much he really misses them. "I started in the egg business twenty years ago when you actually got in a plane and flew to a city and rented a car and drove to see the customer," he says. "Now I have many customers who I actually rarely see, but talk to, like dreams flowing from your brain, via my E-mail system"—which is, he brags, "cable-Internet 500 kbps blazing speed."

For so many people and businesses, speed *is* connectivity. The state of being connected makes them more efficient—maybe even more nimble. Sadly, it also makes them feel busier—maybe even overloaded. If you ran a country law firm out of a gray-shingled building on a local blacktop road, once upon a time your work ended when you had caught up with the morning's mail and prepared the outgoing post. Business ran like correspondence chess, with plenty of time for contemplation. "Unhappily," an American Bar Association pamphlet admonished in 1958, "the public impression is that lawyers are tediously slow." It advised lawyers to adopt modern automatic equipment. Electric typewriters bring one form of pace-quickening ("they are faster" and "require less than one-twentieth the amount of physical energy to operate"). Dictating machines hasten the law firm's heartbeat in another way, allowing the lawyer to work at night and on weekends, without waiting for the stenographer ("Time loss from interruptions is minimized and correspondence is dispatched swiftly. . . . Ideas, time charges, and the like, can be trapped before they leave the mind. . . . Indeed, it can be bluntly stated that, except in rare instances when the secretary's presence in the office at the time of dictation is *absolutely essential,* person-to-person dictation is grossly inefficient and should be eliminated.") We in the era of voice mail and VCR's recognize the benefit as a version of time-shifting. Other professions, other technological speed-ups: medicine has been as profoundly altered by the simple pocket pager as law by the photocopier. Some doctors worry about the rise of what they call "beeper medicine"; they see an addiction to paging and quick fixes. Yes, laboratory results and fresh organs can be rushed to hospital bedsides, and, amid the frenetic twenty-four-hour activity, lives are saved. But the physician risks losing control of his own pace. "All activity becomes crisis oriented," complain Drs. E. Ide Smith and William P. Tunell in Oklahoma City.

> The intensity and zeal of direct paging has now reached such epidemic proportions that newer equipment with storage capacity can take numerous pages simultaneously. The vision of receiving fifty pages per minute becomes a realistic possibility.

In a less connected time, any business deal based on an exchange of paperwork proceeded at a pace controlled by the mails—two, four, six, or more days between volleys. Then came universal overnight mail and its industrial-age children—in Federal Express jargon, "expedited cargo," "just-in-time delivery," "high-speed premium transportation," and "automating and streamlining the supply chain." Federal Express sold its services for "when it absolutely, positively has to be there overnight." In the world before FedEx, when "it" could not absolutely, positively be there overnight, it rarely had to. Now that it can, it must. Overnight mail, like so many of the hastening technologies, gave its first business customers a competitive edge. When everyone adopted overnight mail, equality was restored, and only the universally faster pace remained.

The great instrument of connectedness was, of course, the telephone, transforming the century end to end. Police stations, stock brokerages, and newsrooms managed before telephones but we can barely imagine how. Premodern newspapers sent their reporters to the docks to gather the news from passengers debarking from the great ocean liners. They relied on the mail those ships brought, transoceanic bandwidth measured not in bits per second but in bits per week. News day by day was fast news. The *New York Times* continued its anachronistic Shipping/Mails column as late as 1984. That year, if a law firm had a fax machine, it was an expensive curiosity, perhaps employed mainly for special communications with a particular corporate client. Only eighty thousand fax machines were sold nationwide. Just three years later, in 1987, virtually every American law firm had a fax machine, and within two more years, realtors and takeout restaurants and hardware stores had jumped on the train. Businesses and individual consumers bought two million fax machines in the United States in 1989, and a business card suddenly looked bereft without a fax number. The even faster bloom of E-mail addresses and Web pages, inconceivable at that point, was just six years away.

Connectedness has brought glut. In a group of *n* people, the numbers of possible telephone conversations or dinner-party seating arrangements or sexual-disease transmission vectors grow combinatorially, and combinatorial growth is much faster than geometric growth; it's generally exponential, in fact. Much of the human experience (knowledge, disease) spreads by proximity, and for any one person the number of fellows in proximity has exploded. In past times, even in the most crowded city, we lived close enough to only a few people to, say, read their journals or track the temperature of their hot tubs. Now, in hordes, they put that information on-line. The multiplication of information pathways leads to positive feedback effects in the nature of frenzies. The more people talk and write about the occasional mass phenomena that grab the hysterical attention of American culture—O. J. Simpson, El Niño, Monica Lewinsky, Y2K—the more people want to hear. The more journalists hear, the more they feel able—even obliged—to keep talking and writing. As fluid pressure rises (you learn in high school physics), molecules collide faster and more often, and so the temperature rises too. Close packing and transmission speed are two sides of a coin; that is why sound travels faster through dense crystals. And that is why

Dean Hughson is both rueful victim and cheerful perpetrator of information glut. By 1915, in the fourth decade of commercial telephone service, the American transcontinental system had developed the capacity to handle three simultaneous voice calls. A generation later, AT&T developed a coaxial cable that could handle 480 calls at once. By the 1980s, individual Telstar satellites had enough capacity for nearly 100,000 telephone links, though they were more likely to use the bandwidth for television transmission. Now terabit transmission is coming on-line—one trillion bits per second, or enough for three centuries of a fat daily newspaper. This is the Information Age, which does not always mean information in our brains. We sometimes feel that it means information whistling by our ears at light speed, too fast to be absorbed.

The American company that promoted the Internet hardest in its early days, Sun Microsystems, conducted research in 1997 into how people read on the Web and concluded simply, "They don't." They scan, sampling words and phrases. Why? In part because any one page, on which the fluttering user happens to have lighted momentarily, competes for attention with millions more. Jakob Nielsen, the Sun scientist who carried out the study, cited a typical complaint by a test user, dismayed to be confronted by actual prose—paragraphs of it: "If this happened to me at work, where I get seventy E-mails and fifty voice-mails a day, then that would be the end of it. If it doesn't come right out at me, I'm going to give up on it." Nielsen proposed guidelines for catering to such users—guidelines that came to describe more and more of the Internet reading experience: highlighted keywords, bulleted lists, frequent subheadings, and paragraphs containing exactly one idea. Nothing sticky enough to slow the reader's headlong slide.

Reading E-mail starts to feel like a forced march through a shadeless landscape. More Sun research found, as Nielsen says, that *everybody* who has E-mail complains about the masses of E-mail they get. Interestingly, the complaints are about equally strong no matter how many messages an individual user gets. In other words, people will tell us "I am so overwhelmed: can you believe that I get *ten* E-mails per day" with the same tone of voice as somebody complaining of one hundred messages or more.

My explanation for this phenomenon is that people's expectations for what to do with the mail changes: when they get a little, they treat it as personal correspondence and consider each message and its reply carefully. When they get a lot, most messages immediately are fated for the Delete key. Users are constantly behind on upgrading their behavior on this curve of information neglect, so they constantly feel stressed.

No quills to sharpen, no ink to blot; just bits and more bits, at light speed. Somehow, these same stressed people find minutes to visit a Web site that lets them watch in *real time* what other people are searching for. The search terms flash by, fleeting signposts of information glut: "romantic ideas," "writing AND love AND letters," "cable reel truck," "free clip art," "London real estate," "conduct disorder." It is as if the new World Brain were on display at a science museum and you could peer in and watch the neurons crackling. Technology has opened a direct channel inside. All the stuff pouring in causes congestion, takes up space, reduces productivity, floods the basement, and hyperventilates the attic. That is the sensation, anyway, almost universally shared. More than twenty thousand distinct sites on the World Wide Web address the issue of information overload and, inevitably, contribute to the problem. "Information Overload?" asks an Internet banner advertisement for, it turns out, Microsoft Pointing Devices: the proposed solution is to "Get Moving" with an interactive demonstration of scrolling and zooming.

Who knew that the inconvenience of old-fashioned letter-writing provided a buffer? Highway engineers learned that they could ward off freeway congestion by holding back cars at the entrance ramps, forcing them to wait at seemingly pointless red lights—for their own good, in the long run. In the same way, the unavoidable delays in volleys of business communication before fax, before FedEx, and before E-mail, served as pauses for thought. A lawyer could reconsider a rash piece of mail while it was in the stenographer's out-box. Decisions could ferment during accidental slow periods.

Perhaps we simply have not had time to adjust. We may need to set aside formal time for deliberation, where once we used accidental time. In reaction to the information surplus, a Simplify Your Life movement was born in the nineties. Simplicity loves paradox, unfortunately, and simplification seems to require new fountains of information. For example, Linda Manassee Buell of Arizona, professional coach, trainer, and advisor in life-style development, offered workshops and "teleclasses" on how to simplify your life, plus a 101 Tips booklet, audiotape, and workbook, major credit cards accepted. Macy's advised simplifying your life with the services of its personal shoppers and their "myriad of choices." A drawback of life-simplifying always seemed to be the deprivation required. Some of the things that complicate your life might actually be welcome. "Pretend You Have Just Three Friends," advised *Redbook*. "Stop watching TV news," advised Elaine St. James, one of a cotillion of simplify-your-life authors. "Cancel half your magazine subscriptions." "Cut back on the number of toys you buy your kids." ("When all this is done," one of her readers grumbled, "breathe a deep sigh and say to yourself twenty times, 'I affirm that I have created a life style that does not require my presence.' ") Then again, you might need to get new stuff to simplify your life: the Simply Checking Account; hardwood floors, "voice and fax on demand"; California Closets; a home equity line of credit, Simple® shoes ("single speed"), or any of thousands of other items and services marketed in the name of life simplification.

The whole business turns out to be easier to imagine than to accomplish, and the gurus don't necessarily practice what they preach. Readers have a choice of books offering 100, 52, 365, 99, and 90 ways to simplify their lives. Apparently no author can write just one. St. James alone regurgitates her life-simplification advice in at least five. People mostly read these manuals voyeuristically, the way they read travel magazines with sublime accounts of al fresco meals and blazing white-sand beaches far, far away—meals never to be eaten and beaches never to be visited. The essential simplify-your-life lesson, the idea that launched the phenomenon, is strong and valuable: you have the power to make choices, so make them. Try to distinguish between the little nattering demons that can fill every moment and the greater, quieter spirits that can enrich the passing hours. You may as well, because the life-simplification coaches and trainers will not. They are busy giving birth to an unmistakable Simplify Your Life information glut.

We complain about our oversupply of information. We treasure it nonetheless. We aren't shutting down our E-mail addresses. On the contrary, we're buying pocket computers and cellular modems and mobile phones with tiny message screens to make sure that we can log in from the beaches and mountaintops. These devices are fed by our ever-growing militia of information carriers, professional and amateur journalists. Their spy satellites and listening posts and video cameras ring the globe. Without these information sources we would feel sensory deprivation, as if stripped of our hearing aids and corrective lenses. We can barely understand, omniscient as we are, that the 1941 attack on Pearl Harbor ended an eleven-day voyage by the unseen, unheard Japanese fleet through a data vacuum; or that two thousand people died in the 1815 Battle of New Orleans a fortnight after the relevant peace treaty had been signed in London. We expect information to shine everywhere, soonest. The twenty-four-hour news networks undermine the authority of the traditional network news shows, like it or not—as they, in their time, outgunned the evening papers with a seemingly instantaneous delivery of facts and images. It was not instantaneous, of course, the occasional live feed aside. It was on a time scale of hours; now viewers expect a time scale of minutes. Correspondents who used to scrounge for access to a courier or a Telex machine or, like Peter Jennings, scramble to have film hand-carried from one Asian capital to the next, now carry a complete satellite uplink in their luggage. Along with their laptops and cellular phones, of course.

More than fifty-eight million people in the United States and Europe are "mobile professionals," the Hewlett-Packard Company claimed in 1998, with a need to scan and fax contracts, newspaper articles, and market reports "spontaneously" while they are someplace defined as between other places: driving between sales calls, or on an airplane, or waiting for an airplane. Hewlett-Packard, of course, has new technology to help them capture and transmit this information quickly. We conduct business in bursts. As new items arrive, we curse the offers of FREE 1 yr. USA Magazine Subscriptions and $785,000 Dream Home Give-away!!! We tire of jenny@babeview.com, whose epistolary method is to remark, WoW :{} and See ya, by way of inviting us to look at video footage of naked women. We hear more jokes sliding into the Inbox than we ever did from pals at water coolers; when the Subject line reads, FW: FWD>Fwd- (Fwd) a joke for y, something tells us we're not the first person to read this one, but we do read it, and then we pass it on. Our own little mailing lists of four or six or eight sympathetic souls form tiny enough pathways, yet before long they interconnect globally. Jokes about sex, jokes about UNIX, jokes about lawyers, jokes about *Star Trek* or Bill

Gates in the form of Dr. Seuss doggerel or David Letterman top-ten lists—all these slosh across cyberspace with tidal force. Horribly morbid disaster jokes appear and spread with the kind of timeliness heretofore seen only in tightly knit joke-telling communities of cynical types with access to fast worldwide communication—namely, stockbrokers and journalists. Traveling-salesman humor is obsolete, because we do not need traveling salesmen to carry jokes around. It may soon be a matter of minutes from the time a joke is born to the time every human with a modem has received it.

Every time we curse the overflowing in-box and pass another chain-mail joke along, we expose a disparity between how we feel and how we act. Unless we are masochists and lemmings, we must know something that we aren't telling ourselves. We like the E-mail. We like the connectedness. We do not seem interested in an about-face toward the simpler lives we recall with that rosy, nostalgic glow. Our speedy, in-touch lives can feel good in their own way. The economist Herbert Stein, eyeing the new hordes of men and women who walk city sidewalks with cell phones at their ears and mouths, decided that our need for information on demand is as primitive an instinct as any animal can have.

> It is the way of keeping contact with someone, anyone, who will reassure you that you are not alone. You may think you are checking on your portfolio, but deep down you are checking on your existence. I rarely see people using cell phones on the sidewalk when they are in the company of other people. It is being alone that they cannot stand. And for many people, being alone really means being without Mommy. We are raising a generation that had radio transmitters in its nurseries, keeping Mommy constantly informed of every movement of the baby in his crib. We will soon be walking around with transmitters in our lapels

> or pocketbooks, constantly connected via satellite with Mommy.

A Freudian economist! Their Walkmans, he says, are a way of regaining the steady, comforting beat from the lullabies of infancy. After all, we were born connected. Solitude came with maturity.

Before anyone conceived the idea of bandwidth, before technologists studied information flow as a science, people played chess by mail. In correspondence chess, the transmission of a few useful bytes takes days. The ratiocination is slow, too. This form of chess has now been partly supplanted by on-line competition, but only partly, because some players treasure more than ever the quaint thrill of squandering a hundred hours or more on a single game. Bandwidth in bits per day is a kind of conspicuous consumption or a rebellion against modernity, like wearing spats. In business, not many players can afford quaint thrills. If people want to reach us, we want to be reachable—hence, at the extreme, the Internet fax, the wristwatch pager linked to Web site, and the E-mail system at cable-Internet 500 kbps blazing speed.

Quick—Your Opinion?

CBS News sponsored a real-time telephone poll during George Bush's State of the Union Address in 1992. Nearly twenty-five million attempted calls clogged the nation's telephone network. A bit more than 1 percent of those got through, allowing CBS to tally yes-or-no answers to questions while Dan Rather announced: "Right on to the air within seconds—we've never been able to do that before." Forming an opinion is one process. Stating it is another. Fielding a ground ball is one process. Throwing it to first is another. Sometimes we do best to let one process mature before the next begins.

Opinion pollers invented instant surveys in hopes of removing the burden of time from their work. When they place telephone calls to several thousand people and ask their views of candidates amid a hot campaign, they are trying to hit a moving target. Who knows whether Wednesday's mood has drifted away from Tuesday's? If so, which opinion are the pollers measuring? They also face pressure from clients demanding quick results: opinion as close as possible to real time. Technology has come to their aid. In 1973, the most powerful polling operation in the American economy, Nielsen Media Research, placed new boxes on its subjects' television sets, linked directly by telephone lines to the company's central computers. This system, dubbed "Storage Instantaneous Audimeter," let the company's operations center in Florida track the ebb and flow of viewer behavior minute by minute. That fine temporal grain was a beginning, not an end. Film studios now routinely test their product by letting audiences watch with a dial in hand, registering a sort of instantaneous electric approval. But approval of what?

Along with the language of *real time, time-sharing,* and *multitasking* comes the strange species of bug known as the *race condition.* The race condition is an especially insidious symptom of fast living and sensitive timing. It is what happens when different threads of a program, meant to execute in a fixed sequence, get out of sync. Process A is supposed to create a user account. Process B is supposed to secure it with a password. All the while, files are being created, opened, locked, unlocked, and strange timing errors can occur. A millisecond-wide window may open, through which rogue processes can log in and even gain high-level access. Programmers sometimes assume a lockstep timing that does not exist in a free-wheeling, multitasking world. A procedure vulnerable to an attack of this kind is said to be "raceable." You could race it. It submits to ambiguities of the here and now. By their very nature, race conditions tend to pop up most often in systems trying to perform in real time. A live network handing out tickets to a movie theater, or managing bank accounts, or booking airplane reservations, needs tight control over millisecond-scale operations if it wants to avoid assigning the same seat, or handing out the same dollar, twice.

A race condition without computers is what undermines instant opinion polling. Inconveniently, the opinion being extracted may not yet exist. But the extraction goes on. A tabletop polling dial has become a fashionable gadget at conferences of business executives; speakers keep the audiences involved by soliciting opinions that are instantly tabulated and displayed for the room to see. The Cable News Network has used the same technology to televise instant reactions to presidential debates, showing the results as a live graph broadcast along with the action. Film producers use them, too, in prerelease screenings: audience members offer their real-time reactions in the most herky-jerky, visceral fashion, without even a second's worth of processing or reflection. Was that remark funny or dull? Quick! Is this slow passage perhaps a set-up for something different? We'll never know. *Adagio* is not a permissible tempo for a test audience spinning dials. These are "mood barometers." But moods are smoke in the breeze, and most often these barometers measure something not yet fully formed: an opinion—a *public* opinion—that takes shape over hours or weeks of reflection and discussion. Then again, some instances of public opinion appear faster than ever, as nightly television analysts and daily newspaper commentators compete with Internet pundits to explain events, make judgments, put them into a moral context, and create a kind of instant reflectivity. History cannot really be written that fast. Sometimes events need a decent period of silent mourning. When two young boys stole rifles and shot a dozen of their schoolmates in 1998, the critic George Steiner mocked the instant explanations and the "sound-bite mentality":

> Imagine Dostoyevsky. There are some incidents like this, two boys killing other children, in his famous diary. Imagine what Dostoyevsky would do with that. He would deal with the transcendentally important question of evil in the child.

> Today the editor would say, "Fyodor, tomorrow, please, your piece. Don't tell me you need ten months for thinking. Fyodor, tomorrow!"

By the term "sound bite" Steiner tries to conjure all the quickening of speech associated with television news broadcasts. Sound bites are what politicians learn to speak in, if they wish their voices to be heard in a format that tells whole stories in less than a minute. There are consultants, in fact, who specialize in teaching clients to speak in sound bites: business executives in meetings, authors on promotional tours, and anyone else with a message to get across in a busy, busy world. On network newscasts, sound bites are a distinct enough species to be catalogued and measured. They accelerated from more than forty seconds for presidential candidates in 1968 to less than ten seconds in 1988. Most analysts and many television-news insiders considered that shameful. What kind of depth or insight can a politician aspire to in 8.2 seconds?

But the forty-second sound bite of the previous generation was hardly a Lincoln-Douglas debate. A candidate can dispense glibness and superficiality just as easily in chunks of two hundred words as in twenty. When the networks reacted to criticism and tried to enforce rules against sound bites shorter than, say, thirty seconds, they found their reports growing flaccid. The audience likes the punch of short quotations, and they fit more easily into the structure of a short news report. Perhaps the quickening of video quotation reflects the maturing of the medium as well as the growing sophistication of the audience. "Sound bites have grown short primarily because this medium communicates more effectively that way," argues Mitchell Stephens of New York University. "The fury unleashed by their disappearance is a result of a lack of understanding of and a consequent fear of the new era of video we are entering."

A television news segment approaching three minutes is genuinely "long form," in the context of a half-hour, or twenty-three-minute, network news show. It might be best to think of the one-minute news report as an art form that takes terseness and concision to the limit, kin to the haiku or the oil-paint miniature. You could observe the artists in progress in a typical editing room of the Cable News Network, which spreads its news broadcasting around the clock. The latest editing equipment measures the raw bits of tape in hundredths of a second. "We can put in more subliminal messages that way," says an executive in the doorway. "I'm only kidding." An editor with fast hands on the joystick and keyboard is working with a reporter to tighten a show-business segment, to be on the air in minutes. Psychologists have tried to study the limits of human perception, but these editors, trimming sounds and images to their finest, are practical experts on the subject. If they push their art past a certain point, you will begin to sense fleeting ghosts, present but not quite seen. Push further and you will be lost, no matter how practiced and quick-witted you may be.

"Sound bite" merely names a tool of the trade here—nothing pejorative about it. "We're just doing a sound bite, a sig out, and then we're done," the reporter says. The sig out is the correspondent signing off ("Sam Donaldson ABC News Washington"); the sound bite might also have been called a SOT (sound on tape), and it might have been joined by a NAT (natural sound—horn honking, telephone ringing). They trim a few more words, they watch, and he says: "That's fine. It feels truncated to me. But that's the game."

Meanwhile, public opinion does move at a pace never before seen. Our communal knowledge spreads and assimilates with an almost neuronal instantaneity. Even before internetworking, express mail, and fax, when John F. Kennedy was assassinated in 1963, an estimated 68 percent of the United States population knew within a half-hour. The news of Lincoln's assassination a century earlier took days to spread and sink in that deeply, although railroads and telegraphy were already beginning to form the skeleton of the modern networked world. When George Washington died in Virginia, it took a week for the news to reach New York. In older times, the opinions of masses of people gathered weight on glacial time scales. They were rarely measured or quantified. That made the act of voting in national politics so exceptional. A vote was like the collapse of the waveform in quantum mechanics—a realization of something that was fuzzy and unpredictable until that instant. Imperfect, but at least the act of measurement did not outrace the thing measured.

We are bumping against a speed limit. We can take real-time communication only so far—at least until humanity becomes a single organism with parts conjoined as a light-speed consciousness. The limit is in our own brains. We have finite cruise speeds. "Our own intelligence is tied in with our speed of thought," speculates the cognitive theorist Douglas R. Hofstadter. "If our reflexes had been ten times faster or slower, we might have developed an entirely different set of concepts with which to describe the world."

Decomposition Takes Time

From an article titled "Ten Tall Tales about Composting":

> A number of magazine ads have hoodwinked well-intentioned gardeners into thinking that they must produce compost in fourteen days. Such expectations are unrealistic and unworthy. Decomposition takes time. While producing compost quickly has some merit, no one should feel compelled to purchase chipper-shredders or other elaborate equipment. In fact, even if material *looks* like compost after several weeks, it still requires an additional one-month maturation period before it should be used. . . .

You can't hurry compost for the same reason you can't hurry love and you can't hurry a soufflé. The biochemistry has its own inherent pace. That doesn't mean you won't try.

You suspect, without even thinking about it, that any business called Dombey & Sons, Trujillo & Sons, Eubelhor & Sons, or even Harvey & Daughters must be a venerable business indeed. People are not founding companies today, as they once did, in hopes that their grandchildren will someday carry on the family tradition. No. Grandchildren take time. Nor does one buy deep-blue denim jeans with their dye stiff as tin, resigned to wearing them for a year before achieving a faded "look." One buys them prewashed, prefaded, and maybe prepatched at the knees or seat. Who can wait for nature to take its course? The traditional leather jacket, like a second skin after ten years of wearing, was not actually comfortable in its first year. You had to make an investment. The attachment to old clothes is in the teddy-bear category, growing more from an emotional web of associations than from anything in the cloth. As our reflexes have certain speeds, so does our formation of memories, our accretion of nostalgia. So can the years of breaking-in be effectively bypassed?

Apparently so, because a typical catalogue advertises the "Been There Leather Jacket": "A jacket that (in former lives) has seen it all. . . . There are legends and sagas in each ruck and crease of the distressed, heavy, full-grain cowhide." Naturally it sports "antiqued hardware." In 1997 Disney advertised *The Little Mermaid* as "the timeless classic." Just eight years earlier it was a new movie. Is this, too, a way of straining against the limits of biology—this rapid would-be insertion of a new cultural icon into our store of classic memories?

Modern times have brought certain maladies that might be thought of as diseases of technology: radiation poisoning (Marie Curie's truest legacy); carpal tunnel syndrome (descendant of scrivener's palsy). A unique case is jet lag. This is a disease of clocks. The clocks, of course, are us. Any biochemical process that repeats itself tends to seek a natural rhythm. The rhythms of our bodies—complex, interlocked, and sometimes chaotic—have been entrained by the great astronomical frequencies of our spinning planet. Our bodies know when a day is up, approximately. So it has been through three billion years of evolution. "All the while," notes Arthur Winfree, a specialist in the science of biological time,

> we've felt the sky brighten and darken again and again while the planet relentlessly rotated: a trillion cycles of brightness and dark, of warmth and chill, never missing a beat, deep in the chemical essence of what we are. We are well adapted to the pervasive monotony of sunrise and sunset. . . .

That doesn't mean we have to like the monotony. Our clocks have mechanisms which, it so happens, we can perturb and unsettle. This was difficult to discover. Our internal cycle actually runs closer to twenty-five hours than twenty-four, for reasons that are obscure; it must be reset continually, or we start to drift backward through the hours. We are built, that is, attuned to the time signals beamed from the world around us. As increasingly we flex our muscles and set the pace of our world, so do we remain in sensitive dependence on it.

Through those first trillion cycles, we and our ancestors stayed more or less rooted. The solar time does change, of course, whenever one steps a foot to the east or west. High noon comes a millisecond earlier or later, not enough to notice. With high-speed travel, though—ocean-going ships, railroads, and then, most disconcertingly, motorized flight—came our first serious assaults on the biological clock: dizzying mental paradoxes (just what does happen if you walk around the end of the International Date Line at the pole?) and the malady known as jet lag (the most specialized of the various syndromes that constitute hurry sickness), defined by Winfree as "that disconcerting sensation of time travelers that their organs are strewn across a dozen time zones while their empty skins still forge boldly into the future."

So we tinker with the machine. We shuffle our meals, take

melatonin, or experiment with 10,000-lux light therapy in hopes of shortening the inevitable fatigue and disorientation that comes with long, fast trips across time zones. Maybe we fall for fads involving "biorhythms"—start clocking pseudocycles of twenty-three or thirty-three days, make charts, buy software, check the supposed biorhythms of celebrities. Appropriately enough, Biorhythms is the New Age subcategory just before Divinations & Oracles. But there is a reason for the sudden fascination with biorhythms. Biorhythms matter—biological rhythms that come under stress from, or come into conflict with, tempos that are more easily manipulated. We want to improve ourselves, and sometimes we act gullible. Sony markets a Natural Japanese learning program with the slogan "Learn Japanese in 3 Seconds!" Another company offers a SuperMind Brain Computer: "Put on the light-pulse goggles, and headphones. Push a special button on the command console and an accelerated learning program automatically imprints a complete French lesson onto your brain cells." On good days, we recognize that some biological times simply cannot be reduced below a minimum. Decomposition takes time. Drug testing takes time, because the course of a disease or a pharmaceutical therapy in the human body depends on its complex biochemistry. Even so, the United States Food and Drug Administration accelerated its procedures for approving new drugs in the nineties, under pressure from the industry and without much opposition. The number of drugs approved nearly doubled in 1996 and 1997. A disturbing number of those drugs were recalled soon after. Duract, for example, a painkiller marketed by American Home Products, had the benefit of little more than a year of wide-scale trials; it was pulled from the market eleven months after approval, having caused a series of cases of liver failure and death. Some health-policy experts explicitly blamed the new "fast track" approval process, with its speedy, industry-financed drug reviews.

We humans used to feel like the laggards, with nature marching briskly onward. *Time and tide wait for no man.* "In our day of electric wires," Mark Twain said, prematurely, ". . . we turn it around. Man waits not for time nor tide." Some of biology is essentially a pause: sleep, for example. Pauses serve a purpose, breaking the flow. Like rests in music or caesuras in verse. Like the old nightly break in the news cycle and the financial markets, gone in our 7 × 24 era. Even a confirmed atheist and Sunday driver must believe that the Sabbath served a therapeutic purpose, too, in the epoch when people observed it. Now, of course, Puritanical blue laws ("No woman shall kiss her child on Sabbath or fasting day"!) are mostly long gone, and Federal Express boasts of delivering on Sunday "because the world works seven days a week." Haydn may have been the first great master of the rest in musical composition; he used rests for surprise, rests for tension, and even rests with fermatas. Silence indefinitely prolonged. Rest *and* pause. A rest with a fermata is the moral opposite of the fast-food restaurant with express lane. Modern conservatories find these strangely troubling for some students, who can play the most intricate polyrhythms yet break into a sweat when confronted with this:

Some performers find it difficult to give the rest its full value, let alone the vague extra time called for by the fermata. They just can't wait long enough. There are enforced pauses in the eating of pistachio nuts; preshelled pistachios are an expensive luxury—another *fast* food—and strangely disappointing. It is relevant that researchers in time-compressed speech, discovering hidden punctuation in the pauses that dot our conversation, found that intelligibility declines as the pauses are removed. For most of us, coffee breaks have gone the way of enforced Sabbaths, and neither transcendental meditation nor the sensory-deprivation tank seems likely to replace them.

Biology fights back (or is it technology?). Pauses manage to reinsert themselves into the flow of our faster, multitasking mental lives. Your Web browser is connecting to a distant site, while a voice in the telephone handset at your ear has just said, "An operator will be with you shortly," and you realize three minutes later that you have entered a sort of trance; the operator is *not* with you, and your Web browser has found nothing. Shortly the computer will announce, "The operation timed out." Yes. *Catatonia.* It's the Sabbath.

On Your Mark, Get Set, Think

From a cosmic point of view, the velocity of human thought is more or less fixed—attuned in sometimes useful ways to the velocity of an apple falling from a tree, to the rate of the Earth's spin, to the leaping speed of a predacious coyote, to the gentle passing of the seasons, to the wavelengths of visible light and audible sound. We are defined by these velocities, among others. You could imagine species living on quite different timetables. In fact, you can see them: bumblebees or bristlecone pines, inhabiting temporal planes that barely intersect our own. Careful, though. Speed is not who you are.

From a parochial point of view, we could concern ourselves with small variations in human speeds. We do make a sport out of comparative running speeds. If we were the sort of psychologists who like to be termed "psychometricians," we could pretend to make a science out of comparative thinking speeds. In athletic competition, technology has turned the briefest intervals into arenas for competition. The margins have become so fine that chance easily overcomes the talent that racers strive so hard to perfect. Gusts of wind, uneven turf, random differences in the lengths of swimming-pool lanes can all come into play. The millisecond has come into its own. Baseball, commonly said to be a game of inches, is revealed by the fastest modern cameras to be a game of milliseconds. The pivotal events occur in these tiny windows, testing umpires' reflexes. No ordinary hand-held stopwatch can resolve a millisecond; thus, until recently, a millisecond could not be the margin of victory in sports. Now it can. Luge is one of the events for which Olympic rules now allow a victory by mere milliseconds. Canoeing and bicycling are others. Millisecond sensitivity breeds further dependence on technology. Swimmers clothe themselves in Teflon-coated microfiber suits. Bicyclists ride machines whose aerodynamic properties were honed in wind tunnels. For the sake of fairness, when the world's fastest sprinters line up for the 100-meter dash, the sound of the gun comes to them electronically, to protect against millisecond differences in the time of its arrival through the open breeze. Lasers shine on their backs to provide a continuous Doppler measurement of speed, acceleration, deceleration. And the finish line is monitored by filmless, computer-enhanced, digital cameras, splitting time with a precision beyond the reach of human senses.

Carl Lewis, at his peak, occasionally lost 100-meter races that he had *run* faster than his competitors. His reaction time—the time it takes for the starting signal to translate through eardrum, brain, nerves, and muscles—was generally mediocre, on the order of 140 milliseconds, compared with 115 to 120 milliseconds for the fastest starters. That one-fiftieth of a second difference now matters. It matters so much that reaction times are now regularly monitored. Officials declare a false start not only if a runner moves before the gun but also if a runner moves within a tenth of a second *after* the gun—because reaction times that fast are believed to be humanly impossible.

So runners these days do not just practice running. They practice throwing precisely the optimal piece of torso across the invisible finish line demarcated by the hairline within the electronic camera. They work on their reaction time. They learn to hold a state of intense, hair-trigger alertness, waiting for the signal—but no one, coaches have learned, can hold that state for long, so they hate it when too much time passes after the *set.* "We're talking about thousandths of a second!" says George Dale, president of the International Track and Field Coaches Association. "Flash bulbs, noise from the crowd, a pin drop can make a person move. They concentrate on the sound—that's all they're keyed into."

Can it be that we are finally reaching a point of diminishing returns in racing, a point of virtual perfection? Statistical trends over time suggest that we are, as a species, approaching asymptotically a true maximum speed. Especially in the basic, ancient races between runner and runner, swimmer and swimmer, we may simply be closing in on an absolute limit to the speed that can be drawn in a big and well-trained world from the combination of muscle strength, preparation, and technique. If nothing else, we can no longer tell winners from losers without the aid of a very good clock.

Thinking is not quite so easy to time as running, though. How long did it take an individual human to prove that, for *n* greater than 2, no nontrivial integer solutions exist for the equation $a^n + b^n = c^n$? Years (or centuries, or millennia, depending how one measures). One could make an all-star list of slow but effective thinkers. Charles Darwin considered himself too slow-witted to engage in argument. "I suppose I am a very slow thinker," he said the year he published *The Origin of Species.* Einstein modestly described himself as a slow thinker, but a pathologist kept that famous brain preserved in fluid for years after his death, just in case some future psychometrician could make something of it. Surely in some way Einstein must have been *fast.*

Sometime in the twentieth century the notion started going around that the typical person uses only a small fraction of the brain's true capabilities. You may believe this yourself: *if only we could be trained, optimized, freed to realize our true potential.* "We often hear the cliché, 'We only use ten percent of our brains,' " wrote the popular-culture expert David Feldman, summarizing a question that readers sent him. "How is it determined that we use ten percent and not five percent or fifteen percent?" The idea seems to have bloomed at Ohio State University in the 1940s. A Gestalt psychologist there, Samuel Renshaw, claimed to have demonstrated that the average person achieves "on the order of twenty percentile utilization of the sense modalities." The *Saturday Evening Post,* publicizing Renshaw's work in 1948, translated this as, "Most people are only about 20 percent alive. . . . We use our eyes, ears, noses, taste buds, sense of touch—and minds—at one-fifth or less of potential capacity." One of Renshaw's specialties was speed-reading. He claimed to teach college students to read five times faster than before—more than a thousand words per minute. Speed was, in fact, the essence of his method. He used a mechanical tachistoscope, or "glorified magic lantern," to project images onto a screen for as little as one-hundredth of a second. His subjects could learn to see and read back short strings of numerals flashed in front of them, and the Navy hired him during World War II to quick-train airplane spotters. With practice, people could learn to feel right at home in these tiny regions of time.

The idea of huge portions of our brains lying dormant does not bear up well under scrutiny as a biological fact. Really—all those uncounted neurons waiting idly for modern schooling? Maybe it is a form of cultural truth, though. The notion seems to reflect an altered sense of the relationship between humans and the world passing before our senses. By the twentieth century, that world had accelerated so much that we had to feel changed ourselves. Our ancestors did not have occasion to process tens of thousands of words a day, spoken or written, any more than they challenged their visual cortex to process cinematic montage. Even without Renshaw's tachistoscope, people unconsciously sensed a growing fulfillment of human potential. Renshaw was not so much training people to perform great mental feats as he was discovering that people do perform great mental feats. We absorb the stimuli of a fast, complex world; we require specialized training to handle parts of it (automobiles, VCR's, Gameboys); and often enough we enjoy it.

We don't automatically admire quick thinking—it can mean glibness and superficiality, we may have noticed. But we do associate it with intelligence, these days more than ever. Lightning calculators must surely be smart, the occasional *idiot savant* notwithstanding. Quick-witted people, the mentally agile, those who can think on their feet—we may not always choose them to be captain or president, but we tend to respect them. We have heard of unhurried qualities like wisdom and sagacity, but we think nonetheless that the students who plod through laborious calculations cannot be quite as smart as their comrades who snap their fingers and know the answer. Some modern businesses have built this assumption into their hiring procedures. "At least in industries like high-tech and finance, quick-wittedness rules," observes Nicholas Lemann, author of *The Big Test.* "Some companies, such as Microsoft or D. E. Shaw, the stock-picking firm, are particularly known for hiring on the basis of mind-speed and for peppering job applicants with SAT-like questions in interviews so as to bring the quality into high relief." Much of life has become a game show, our fingers perpetually poised above the buzzer. We're either the quick or the dead. To be *quick* it used to be enough merely to be alive. Now we expect repartee and fast response times, too.

"This is not a state of affairs that would, at most times and places in history, have been considered normal and healthy," Lemann adds. The expectation has impatience as its corollary. For every high-school student running out of time as he finishes his SAT's, for every Microsoft job applicant asked to solve a logic puzzle in a given number of seconds, there is a judge waiting at the other end, drumming imaginary fingers on an imaginary desk. Strange though it may seem now, most of the brief history of intelligence testing was more patient. The first psychometricians, eager as they were to find some real, innate, general quality of mental ability that they could measure with tests, rarely paid attention to *speed* of thought. Some tests had time limits, others did not, and it was far from clear that success on the former, "speeded," tests had any more to do with intelligence than success on unspeeded tests. Charles Spearman, the British psychologist who invented the notion of a general intelligence called *g*, was not thinking in terms of quickness. He imagined the physical basis of *g* to be a sort of energy—a natural thought for a scientist of the early 1900s, when physicists were discovering many strange new forms of energy, visible and invisible. Underlying intelligence, he argued, must be "something in the nature of an 'energy' or 'power' which serves in common the whole cortex (or possibly, even, the whole nervous system)." Perhaps this energy fuels groups of neurons acting as "engines." He was admittedly speculating: "There seem to be grounds for hoping that a material energy of the kind required by psychologists will some day actually be discovered."

Half a century later, no psychometrician had been able to find differences in brain energy that corresponded in any interesting way with intelligence. Meanwhile, though, computers had arrived on the scene, providing a fertile source of new metaphors for brain function. Just as PC buyers grew obsessed with benchmarks of processor speed—4.77 megahertz (million cycles per second), 6 megahertz, 16, 20, 66, 100, 233, 300, 550, and rising (exponentially, of course)—a group of mostly American psychologists began searching for measures of the information-processing speed of the brain. Never mind that carbon-based brains do not run on the digital clock cycles of their young silicon counterparts. Beginning in the 1970s, psychologists attached electrodes to subjects' arms or skulls to measure nerve-conduction velocity, the speed of electrical impulses through the nervous system. You may be disappointed to learn that yours is only about 160 feet per second. They measured reaction time, "RT," the time needed to recognize and act on a stimulus. In tests devised by the Berkeley psychologist Arthur Jensen, subjects had to move a finger to whichever of eight buttons had just lit up. This measured more or less the same quality that separates the gold medalist in the 50-meter freestyle from the silver medalist who dived a tenth of a second too late. Jensen, however, claimed much more for it: "University students show faster reaction time than vocational college students, who are in turn faster than unskilled factory workers, who are faster than the mentally retarded," he wrote in 1984. He and his colleagues looked for other "chronometrics," measures that could be allied with some biological version of processing speed. There is inspection time, often measured with tachistoscopes and their descendants, by flashing a pair of lines on a screen for a fraction of a second and asking which was longer. There is the time it takes to decide whether two words are the same or different and the time it takes to recall whether a number was in a string of digits previously seen. Many researchers now claim to find correlations between these different measures of brain speed and the results of intelligence tests. "The simplest interpretation of these results is that intelligent brains are faster," asserts one.

On closer scrutiny, interpretation is not so simple. In the notoriously politicized field of psychometrics, researchers have a tendency to overlook experiments with zero or negative correlations, and there are many. Sometimes the results of these different speed measures do not even correlate well with one another, let alone with the results of IQ tests. Often, fast reaction times seem to go with success on unspeeded tests but not on speeded tests, a problem for advocates of the fast-processor view of intelligence. It seems intuitively plausible that fast, efficient neurons would be useful for anyone looking for work as a software engineer, but our processors are still more complex than Intel's. One quality the psychometricians are clearly testing is the ability to take tests—to concentrate on tests, to devise strategies for tests, to learn the perceptual patterns of tests. Pared of sociological baggage, the real lesson of the research is that people who do well on psychologists' tests do well on psychologists' tests. Usually. "If anything," says the Yale psychologist Robert J. Sternberg, "the essence of intelligence would seem to be in knowing when to think and act quickly, and knowing when to think and act slowly." Stop a moment and ponder.

A Millisecond Here, a Millisecond There

Between thoughts, there are gaps—very, very short gaps. Can this time be used?

Measurable breaks separate the songs on record albums. Some are longer than others. That is usually deliberate. A sensitive record producer will run songs almost together or leave a perceptible pause, depending on the desired effect. Someone, though, must have realized that these gaps are a waste of time, in the same peculiar sense as the momentary fadeouts between segments of television programming. It had to be someone who realized just how long a second can be—not a mere instant anymore, but a space stretching before us as a hectic container, with events and voids, to be filled with milli-, nano-, or picothings. Certainly a second is long enough for impatience to begin welling up. So the Sony Discman, circa 1996, offers a function that lets the user close the gaps on CD's. The instructions suggest, "You can enjoy playing with less blank space between the tracks."

The evolution of technology has long been about saving time, but on grosser scales than now. Certainly the cotton gin, the automobile, and the vacuum cleaner let people work, move, and clean faster—savings to be measured in hours and minutes. Now we're saving fractional seconds: a millisecond here, a millisecond there—does it really add up? The consumer-product laboratories think so. They are slicing time ever more finely for us. Other kinds of inventors may be making more profound use of their windows onto the millisecond world. Air bags, as a life-saving feature of automobiles, were conceived and designed only when it became possible to visualize complex mechanical sagas happening—beginning, middle, and end—in one-tenth of a second. The creators of air bags were carrying on in the trail blazed by Muybridge and Edgerton. So why not use some of that new knowledge of time's microcosm to help out in daily life? Toasters are toasting faster—pushing the limits set by the thermal conductivity of bread, if you want the center warm before the surface blackens. It could take two or three minutes for an under-the-tongue thermometer to rise to your temperature; new thermometers are electronic and, naturally, faster. By comparison, the time-saving promised by J. F. Lazartigue's *séchage rapide* shampoo seems gross and vague: its polymers with perfluorides purport to hasten drying by 30 percent. The household-products designers at companies like Black & Decker, developers of the Dustbuster miniature vacuum cleaner, find time-saving opportunities all through the household day. Owners of a Dustbuster need not waste time walking to the closet, finding an outlet for the power cord, or rewinding the power cord. They may buy extra Dustbusters to be spread strategically around the house. There are still seconds wasted in ironing—the heat-up time—which the Black & Decker people have plucked with their new HandyXpress iron, for the "hurry-up market." They cite Gallup survey research to the effect that a majority of Americans, and especially baby-boomers, feel that they "do not have time to do everything that needs to be done." The answer may be self-evident; the question, surely, is revealing.

Computer printers have saved time in a dramatically accelerating rush of their own. There used to be people known as typists; before that, clerks, copyists, and scribes. The labor of imprinting words onto paper consumed at its peak a staggering share of the economy's time budget. Herman Melville's fictional scrivener Bartleby set down text for lawyers at the standard rate in the mid-nineteenth century of four cents per hundred words. In that world the coming of the typewriter created a revolution as fierce as the introduction of moveable type into a world of monks and quills. Mark Twain, who acquired one of the world's first typewriters from Remington, instantly admired its word-slinging velocity. He typed to his brother:

> I am trying t to get the hang of this new f fangled writing machine . . . The machine has several virtues I believe it will print faster than I can write. . . . It piles an awful stack of words on one page.

Faster than a writer could write—thus fast enough, surely. This breakthrough machine sped the pace of nineteenth-century business; not incidentally, it brought a wave of female typists into male workplaces; it "carried the Gutenberg technology," as Marshall McLuhan said, "into every nook and cranny of our culture and economy."

Yet it did not take long for typing to seem slow. An author finishing a book or a student finishing a long paper had to set aside days or weeks for the typing of the final draft. When mass-market computer printers arrived, in the 1970s and 1980s, they brought a new metric: instead of words per minute, characters per second. A good printer, its hard-alloy daisy wheel clattering across the page, could spray twenty or forty or even eighty characters per second, like paint from a magical can. A page every minute, and the output was almost as crisp as that of an IBM Selectric typewriter. For an author, who could finish a book and see it in final typescript *the same day*—a day of sitting beside the printer and feeding in paper, sheet after sheet—the effect was inspiring. And forgettable. Within a decade the same author would likely be exasperated with the wait for each page from a much faster printer. The daisy wheel was already as obsolete as the Selectric, and for the same reason. Too slow. Printers, like so many devices of the modern world, had just one essential measure, speed. Instead of characters per second, pages per minute. How many, ideally? Infinite would be about right. "The ideal printer," explains an encyclopedia of computing, "would be one that produces its work as soon as you give the 'print' command, all fifty thousand pages of your monthly report in one big belch." We're not quite there, but already, in the celestial calculus of saved time, this progression of technologies—typewriter, electronic typewriter, line printer, page printer—has racked up an astounding tally of days, hours, and minutes. Where did all this saved time go?

Our world handles time in smaller and smaller coinage. Second-saving technologies can be simple—quicker-heating coils in toasters and irons—or clever. The portable CD players use memory chips to store a few seconds of music and feed it back—recovering from playback errors and squeezing those blank intervals as a side benefit. Some new telephone-answering machines have quick-playback buttons. These are for handling callers who have droned on and on with their shaggy-dog messages. Because the technology is digital, the pitch doesn't rise à la Chipmunks; the sound just goes by faster. How did the manufacturers know you were so busy that you could not stand to listen to your friends speak with normal languor? It's no secret. You like them faster. The current generation of answering machines seems to favor a 25 percent speed-up. Perhaps we will soon learn to expect and understand speech that is even more rapid. As we surround ourselves with these quick technologies, we sometimes begin to doubt ourselves. We measure ourselves against our machines, and we worry that we are lagging behind. They are faster than we are. A poor human can't keep up. Then again, we can. These are tools, not competitors. Even computers, terrifyingly speedy as they are, keep us waiting, we may note smugly.

Who can say just where we began the slide down this long, strange slope of milliseconds? One place may have been the New York World's Fair in 1964. Many thousands of people waited in line at the AT&T pavilion to try out Touch Tone dialing for the first time. Robert Moses, as fair president, got one installed at his desk, so that he could gain a tempo on the millions of New Yorkers still spending (the company estimated) an average of ten seconds per seven-digit number. Visitors got to dial numbers the old rotary way and then the new push-button way, for purposes of comparison. An electrical readout showed just how many tenths of a second they could bank. However many it was, we now know it wasn't enough. Subsecond time-saving was already a telephone-company tradition. When the century began, time-saving inspired the Bell mentality. "The telephone saves time," began a 1904 company credo. "It saves time in business where time is money. It is to make the saving of time as great as possible that the Bell Telephone Companies are constantly trying to save a fraction of a second here and there." The spirit was contagious. In the post–Touch Tone generation, you probably have speed-dial buttons on your telephone. Investing a half-hour in learning to program them is like advancing a hundred dollars to buy a year's supply of light bulbs at a penny discount. Meanwhile, in some places, telephone directory assistance now offers callers the option of automatically dialing the number they just retrieved, for a price. A case study in what time is worth: in the New York area, soon after this service began, 21 percent of customers proved willing to pay 35 cents to save about two seconds. Add them to the ledger.

1,440 Minutes a Day

About that ledger. *Where does the time really go?* Start with the usual twenty-four hours. You spend seven hours and eighteen minutes asleep, on average. If you believe the statisticians.

By the way, this is not enough. Clocks having replaced the natural rhythms of light and dark, researchers believe that people need to sleep much longer than they do: at least eight and a half hours. But it is hard to get things accomplished in this condition. No wonder marketers try to sell tapes promising to help you make money while you sleep, burn fat while you sleep, or learn foreign languages while you sleep. Set up your computer properly and you can at least download megabytes from the Internet while you sleep. Still, the statisticians claim that you are going to bed too late and waking up too early, saving more than an hour every day for more interesting activities and paying the price in terms of a broad societal trend toward sleep deprivation—hypersomnolence, sleep apnea, and all-round fatigue and exhaustion. The National Sleep Foundation estimates that average sleep time has dropped 20 percent over the past century. The mere presence of an alarm clock implies sleep deprivation, and what bedroom lacks an alarm clock? According to a persistent legend of the early 1960s, a so-called Russian sleep machine could save a person six hours a day by sending a gentle, sleep-intensifying electric current through the brain. "In two hours the brain's owner has had a full night's sleep," *The New Yorker* reported dubiously in 1963. "Up he pops, we must suppose, crackling and full of beans, all cobwebs gone, at two in the morning. . . . Look at it from either end, he has saved a whole lot of time." If only he could. Instead, sleep-disorder clinics have more than tripled in the United States over the past decade. We are a collection of zombies, researchers believe; sleep loss is epidemic. And possibly contagious—are you keeping a spouse awake, by any chance? All right, maybe you actually treasure your waking time in the dark: the nightclubs, or modern equivalent; the restless sounds of city streets and people in their own state of insomnia; the chance to buy a bagel and fresh coffee and tomorrow morning's newspaper at midnight. That's you. No one can measure the global cost of sleeplessness: lost alertness at work, clouded thinking at air-traffic centers, a large fraction of all car and truck accidents, and distraction in the control room at Three Mile Island. In our groggy condition, the slicing away of an hour each spring at the onset of daylight saving time—our once-a-year twenty-three-hour day—causes a noticeable rise in car crashes and accidental deaths of all kinds. A Canadian psychologist, Stanley Coren, puts this increase at 6 percent. The sleep-deprived begin to surrender willy-nilly to momentary "microsleeps," he says. "Eventually, if the sleep debt becomes large enough, we become slow, clumsy, stupid, and, possibly, dead."

Anyway, you sleep. This leaves you with a bit less than seventeen waking hours.

You spend one hour and thirteen minutes each day driving a car, mainly just to get to work, or so reckons the Federal Highway Administration. From 1970 to 1996 the mileage driven by Americans rose four times as fast as the population and eighteen times as fast as the number of new roads, the Federal Highway Administration reports, with the all-too-predictable result that driving has become one of the few mainstays of modern life that are genuinely slowing down. Average speeds at rush hour have declined, reaching fifteen to twenty miles per hour. We can only marvel at the growing dissonance between our streamlined highway technology and the reality of traffic. We can marvel in real time, as traffic reports are beamed to drivers via electronic highway signs or cellular phones and pagers. In this case knowledge is not power. "Freeway systems are designed with generous high-speed lane widths, long acceleration and deceleration lanes, pull-off shoulders, and superelevation on curves, as well as long sight lines and no cross traffic," notes a transportation policy analyst, Peter Samuel. "Superhighways are set up for our cars and trucks to cruise along safely at 50 to 70 mph, yet they are becoming parking lots full of stop-and-go traffic for hours each day."

Like the ticket lines for central cultural events (World Series games, pop-superstar concerts), like the routers and switches of Internet gateways, like National Park campgrounds and the White House E-mail, the world's urban roadways have become excellent places to measure congestion. Perhaps the most disciplined congestion research in recent years is the Texas Transportation Institute's Urban Mobility Study, covering more than fifty American cities. The Texans are serious. They have examined lane-mile by lane-mile of freeways and arterial streets, applying their formula for congestion:

$$\text{Roadway Congestion Index} = \frac{\frac{\text{Freeway}}{\text{VMT/Ln.} - \text{Mi.}} \times \frac{\text{Freeway}}{\text{VMT}} + \frac{\text{Prin Art Str}}{\text{VMT/Ln.} - \text{Mi.}} \times \frac{\text{Prin Art Str}}{\text{VMT}}}{13{,}000 \times \frac{\text{Freeway}}{\text{VMT}} + 5{,}000 \times \frac{\text{Prin Art Str}}{\text{VMT}}} \qquad \text{Eq. S-1}$$

They find that in Los Angeles alone, more than 2.3 million person-hours were lost to traffic delay in 1994. There, in the land of the seven-digit license plate and the ten-lane freeway, traffic has roughly tripled since 1970, whether measured in lost person-hours or miles of congestion or vehicle-hours of lost time. Researchers predict that average speeds will continue to fall, to 11 miles per hour by the year 2010. In New York, where 11 miles per hour might not sound so bad, more than 2.1 million person-hours were lost to traffic in 1994. These hours amount to dozens of new life prison sentences. As drivers know, it can be a peculiarly infuriating and claustrophobic prison. The dynamics, studied as one would study the dynamics of any complex system, prove strange indeed. On highways, every driver has seen traffic appear mysteriously, jamming the flow as if something were blocking a lane up ahead, and then dissipate just as mysteriously. "I actually have it on film," said Stephen L. Cohen, a mathematician with the Traffic Systems Division of the Federal Highway Administration, after monitoring I-95 in Virginia. "You have a flow close to capacity, and one or two guys slow down, and everybody behind them starts to slow down, and it sets up a shock wave. It can be deadly." Sometimes the congestion is the secondary shock from an accident that blocked traffic minutes or hours before. The wave of congestion moves upstream more rapidly than the congestion clears later, when the actual obstruction is gone. So even after nothing remains for the rubberneckers, elsewhere the flow preserves a kind of memory of the incident, still blocking cars far from the actual site. At other times, the waves of stop-and-go traffic are simply the consequence of flow too close to the critical point of saturation, where it becomes sensitive to the smallest perturbation—a single driver stepping on the brakes. Stop-and-go driving creates its own special instabilities. These are the behaviors of cars in herds—choking masses of congestion that rarely appeared in a slower era. Meanwhile, in the midtown core of New York City, at any given moment, astonishingly few vehicles are in motion: about nine thousand. Add a few cars to that number, and the flow does not just slow. It gels. It creates the famous condition dubbed "gridlock" by New York traffic engineers of the 1970s. Technically speaking, gridlock is not, as many believe, the freezing of a single intersection by traffic entering from two directions. That is "spillback." Gridlock is the even more devilish condition that occurs when traffic jams up all the way around a city block and becomes a circular cascade of congestion, so that every car is actually blocking itself, the snake biting its own tail. Another positive-feedback effect. Another paradox for hurried times.

Per "eligible driver," the Texas research measured annual hours of delay in the dozens for virtually every sizeable city. Washington, D.C., ranked first: seventy-one hours. As the research continued through the nineties, congestion continued to worsen. Generally, the drivers, wondering how late they will be or trying to compute alternative routes or just pounding on their steering wheels out of habit, find it difficult to relax.

Assuming a minimum of overlap between sleep time and drive time, that leaves just over fifteen hours out of bed and out of car. What to do . . .

Sex and Paperwork

Americans tell pollers that their single favorite activity is sex. In terms of enjoyability, they rank sex ahead of sports, fishing, bar-hopping, hugging and kissing, talking with the family, eating, watching television, going on trips, planning trips, gardening, bathing, shopping, dressing, housework, dishwashing, laundry, visiting the dentist, and getting the car repaired. Yet who has time for it? The broadest and most careful modern survey of American sexual behavior, conducted in 1994 by a team from the University of Chicago, suggests that the average time per day devoted to sex is four minutes and a few seconds. That is, on average, one half-hour event per week. Not much—even if this four minutes excludes time spent flirting, dancing, ogling, cruising the boulevards, toning up in gyms, toning up in beauty parlors, rehearsing pickup lines, showering, thinking about sex, reading about sex, doodling pornographically, looking at erotic magazines, renting videos, dreaming of sex, looking at fashion magazines, cleaning up after sex, coping with the consequences of sex, building towers and obelisks, or otherwise repressing, transferring, and sublimating. Thank heaven for quick-release fasteners. Do you find the half-hour figure implausible? Your own sexual time budget is much larger, of course—or much smaller.

When you add up the minutes, they seem dismally few for the activity which, above all others, is supposed to stop time, destroy time, and lift us out of time. "Ye gods!" wrote Alexander Pope, "annihilate but space and time, and make two lovers happy." No wonder time-use researchers, looking in their subjects' time records for evidence of any sexual activity at all, reckon that sex must be slipping in as miscellaneous "free-time hours" or "general personal care." Four minutes a day? Ye gods!

Another four minutes a day goes, on average, to filling out paperwork for the United States Government—the paperwork subject to the Paperwork Reduction Act of 1980. This historic legislation marked a new official recognition of a fact as old as the phrase "red tape": that government taxes its citizenry in a currency of minutes as well as dollars. By 1980 the minutes mattered enough to justify a huge new enforcement apparatus within the federal establishment. Four minutes a day is the official number, even if the government is not fully at ease with its own arithmetic. "There is a total, which I will disavow as soon as I say it," explains Sally Katzen, the federal official responsible for enforcement of the act. "Six point nine billion hours. It sounds enormous, and it is indeed a big number." That total, from 1995, includes voluntary and semi-voluntary paperwork: customer-satisfaction forms at National Parks and passport applications. It includes paperwork required for receiving benefits. It does not include paperwork by government agencies themselves—an exception that conveniently breaks an infinite loop. "The Paperwork Reduction Act does not have to comply with the Paperwork Reduction Act,"

Katzen notes. The act does, in fact, inspire its own paperwork, as befits a multifaceted society, but it also succeeds in damming some of the paperwork tide. Every year, time is saved—for example, 130,000 hours that had been spent by anglers who might have accidentally killed or maimed marine mammals; whereas formerly they were required to maintain a daily log, now their paperwork is confined to reports on actual incidents. As an average member of the United States population, you save one two-hundredth of a second every day by not having to keep that particular log.

Most of all, government paperwork in the United States means the filling in of tax forms. Each form comes with its own set of time-use numbers, magical-seeming in their precision. For example, suppose you are lucky enough and scrupulous enough to be filing Form 730, Tax on Wagering. It is a simple form, just one short page plus instructions. By law, the instructions must include, prominently, the following:

> The time needed to complete and file this form will vary depending on individual circumstances. The estimated average time is: **Recordkeeping,** 3 hr., 35 min.; **Learning about the law or the form,** 1 hr., 10 min.; **Preparing the form,** 2 hr., 12 min.; **Copying, assembling, and sending the form to the IRS,** 16 min.

Seven hours. The Office of Management and Budget had to ensure that this burden bears a tolerable relationship to the legitimate needs of the Internal Revenue Service. But the enforcers of the Paperwork Reduction Act do not have the power to ask the tax collectors to give up on collecting particular taxes, and someone accepting a seven-hour paperwork burden in order to pay a wagering tax of a few dollars may wonder whether the effort is worthwhile.

Time spent on paperwork is part of the cost of living in a complex world. Your mix of activities and responsibilities becomes a machine with more and more interlocking parts, and the cost of running it rises nonlinearly. Do you have a car? Then, do you keep the warranty paperwork in order? Carefully file your repair receipts? Do you sort out the credits and riders that bulk out the documentation of your insurance coverage? Keep a logbook to account for business and personal use when tax time comes around? Would you like to qualify for an on-off button to control your passenger-side air bag? (If so, you and your physician will both spend time on intricate paperwork.)

At least the Paperwork Reduction Act created an opportunity to measure some of the time sucked away by our dealings with the federal government. No one can even guess at the world's time burden for person-to-person and person-to-business and business-to-business paperwork. Some part of that comes with paying bills. Most people spend no more than a minute or two a day on this chore, yet it weighs disproportionately on the mind as, by definition, a draining obligation. Perhaps you have begun to speed up your personal finance with a personal-computer check-writing, check-printing, checkbook-balancing, portfolio-managing, bill-paying product. The marketers responsible for the most popular bill-paying software minced no words about its main selling point when they named their product "Quicken." It has a "streamlined interface." It has "OneClick Shortcut Technology." It has QuickTabs, QuickZoom, and QuickFill: by the time you have typed "b-l-o" the software has guessed that this is your Bloomingdale's bill—another fraction of a second saved. But does such software save time overall? That may depend on whether you are the sort of person who can be sucked into creating color pie charts to break down your grocery budget in fine detail. To save time, you must invest time. Do you succumb to the automated plea that you "register the software" (an industry euphemism;

what you really register is yourself) and discover, after filling in the on-screen forms, that you need to set up your modem yet again? (And where is the manual for that?) Certainly you will want to create separate accounts for checking, saving, credit cards, loans, and securities. If you enter categories for each transaction, you can quickly generate reports. With bar graphs! You can print checks, with a professional look, more quickly than you could have written them by hand, especially if you order the special laser-printer checks. If you enter advance dates in the built-in calendar, the program will remind you when it is time to pay bills. If you record your assets and liabilities, you can assess your net worth. If you enter your purchases of stocks and bonds, you can track the performance of your portfolio. You can plan for retirement and for children's education. So just think of all the time you are saving, compared to the days when you balanced your wallet-size checkbook by hand. And when tax time comes you will be ready.

Those exceedingly precise average times listed on every tax form trace their ancestry to a giant study performed in the 1980s by the Arthur D. Little research organization. It was the first comprehensive effort to measure what the researchers defined as "the time-component of burden measured in units of human hours of effort." Nielsen ratings, meet Form 1040EZ. There were questionnaires, focus groups, and diaries. The Little analysts considered hundreds of variables affecting time drain, from *items on a form requiring records* to *total number of words of instruction.* They tried to take into account every conceivable sinkhole for taxpayer time: *saving, sorting, or filing receipts*; *telephoning, writing, or making trips to get records of expenses and income, such as doctors' receipts, W-2 forms, etc.*, or *to get tax forms*; *reading books, magazines, or articles about taxes*; *making calculations*; *using a computer to do any of the above.* The resulting mathematical model, delivered to the government in the form of an elephantine computer program, is run and rerun every time the IRS changes a form. Is it realistic? Best not to waste time worrying. By its own admission, the Arthur D. Little study had to exclude "psychological costs."

So, sex and government paperwork: four minutes each. That still leaves about fifteen hours to spend as you will.

Whether or not you have a child, you spend thirty-one minutes on child care (it's an average). Likewise, your time budget includes seven minutes devoted to the care of your plants and pets, whether or not you have any.

You spend sixteen minutes looking for lost objects (a year of your life), or so *American Demographics* reported.

You spend twenty-nine minutes visiting other people—a number that has declined dramatically over the decades. Telephoning is faster than walking. Face-to-face communication takes time to establish, you might say.

To compensate, you spend fifty-two minutes talking on the phone—a 1990 figure, up from thirty-six and a half minutes a decade earlier. Add to that, in the business world, an average of fifteen minutes a day spent on hold.

As for time on-line—time getting on-line, time waiting on-line, time actually browsing, chatting, or surfing on-line—estimates are obsolete as soon as they are formed. For most people, the number is still zero, but one survey found users spending eighty-six minutes a day. Teenagers are supposed to be the heavy users, but market research finds people aged forty-five and up sitting at their computers for more than seventy minutes a day. Another study finds that Internet users spend nine painful minutes a day just waiting for Web pages to materialize on their screens—never mind time spent actually reading these pages.

By one "conservative" estimate, an active user spends almost four minutes a day just booting up and shutting down his personal computer. Thus Windows 98 could be marketed as a time-saving technology. "I saw Bill Gates demonstrate faster boot-up, application loading, and shutdown," reported John Dodge of *PC*

Week. "I got excited. If I recoup just thirteen minutes a week, I've won back 23.5 days over fifty years of PC use." No wonder Ivan Seidenberg, president of Bell Atlantic, jokes about the DayDoubler program all his customers seem to want:

> Using sophisticated time mapping and compression techniques to double the number of hours in the day, DayDoubler gives you access to 48 hours each and every day. . . . At the higher numbers DayDoubler becomes less stable, and you run the risk of a temporal crash in which everything from the beginning of time to the present could crash down around you, sucking you into a suspended time zone.

And the average American man spends more than ten minutes a day shaving, if you believe the Schick Razor Company. Then he spends almost a minute choosing and tying a necktie. If he is too young to shave, he spends "almost half an hour" coloring with crayons, according to Crayola.

Modern Conveniences

Four minutes is also the amount of time a microwave oven *saves* you each day, if you are a woman eighteen to fifty years old. Without the microwave, you spend fifty-five minutes preparing food; with it, fifty-one. This does not include time spent buying the microwave, cleaning it, maintaining it, and feeling guilty about not reading the instructions for programming it. Strange that the time savings are not greater. The microwave oven is one of the modern objects that convey the most elemental feeling of power over the passing seconds. You watch those seconds, after all, as they tick past on the digital display. If you suffer from hurry sickness in its most advanced stages, you may find yourself punching eighty-eight seconds instead of ninety because it is faster to tap the same digit twice. You face new dilemmas: does standing at the microwave for a minute and a half make you feel that you are wasting time? Will you be able to apply these time savings to your chores, your obligations, your assignments? "Ah, let them go," Randall Jarrell wrote, while *not* looking through the window of a microwave;

> you needn't mind.
> The soul has no assignments, neither cooks
> Nor referees: it wastes its time.
> It wastes its time.

Fine, easy for him to say, but is a minute and a half long enough for the soul to make a quick phone call or run to the next room? It's surprising what can be packed into eighty-eight seconds. If you just stay and stare through the oven window, time leaps forward before your eyes. You see the food coddle or steam preternaturally fast. You remember a Steven Wright joke: "I put instant coffee in my microwave oven and almost went back in time."

Other time-saving appliances, too, seem to accomplish disappointingly little in the way of creating leisure. A dishwasher saves barely one minute in clean-up time. According to the industry, this is because people needlessly scour the dishes before placing them in the machine. Or they take advantage of the convenience to use more dishes. All in all, the average woman still spends almost four and a half hours a day on housework, at least according to an often-cited 1987 survey. That is more than twice as much as men, notoriously. Men try to compensate—the little dears—by spending twelve minutes a day on outdoor chores and sixteen minutes a day on home repair, not even counting time spent researching these activities via *This Old House*, *Home Again*, or *Home Improvement*. All these housework numbers have fallen over the decades. Researchers suspect that dust is collecting in places that might have been wiped clean in an earlier generation.

By the way, maybe you, defying the averages, do read the instructions in the owner's manuals for your appliances. If so, you accept another of the staggering time burdens that come with being a good citizen of an intricate era. (If you don't, you may occasionally feel guilty about it.) *Read and understand all instructions,* the typical manual begins. *Observe all warnings and instructions marked on the product.* Plan the location of power cords carefully, and make sure the voltages remain within limits. *If you are not sure of the power supply to your home, consult your local power company.* If you have trouble with the polarized two-prong plug or the three-prong plug for grounded outlets, consult a licensed electrician. Many appliances urge weekly cleaning, inside and out—check the manual for the full program. *Clean exposed parts with a soft, damp cloth.* Then clean the cloth. By now you have established a storage and filing system for these manuals and associated materials. *Save the original packaging to protect the system. Remember to save your sales receipt in case you ever need warranty service.* You can't be too young to spend time on these matters; the ubiquitous Furby doll of 1998 came with lengthy "Before You Play with Me" Instructions: . . . *3. While holding me upside down, put your finger in my mouth and hold down my tongue switch, then hit Reset button. . . .*

For your further convenience, many of your time-saving devices employ batteries. Surely you have developed a certain expertise in caring for these. You know enough not to recharge a cellular-phone battery before it is fully drained of power; or to carry AA batteries loose in your pocket, where contact with keys might cause them to short, leak, or rupture; or to store batteries in damp places and at abnormal temperatures. You are diligent in removing every battery from toys and cameras when they are not in active use, lest you risk corrosion and leakage. You have educated yourself about the difference between nickel cadmium and nickel metal hydride.

The battery, this most mundane of objects, has become another of the hidden time-swallowers in the everyday lives of consumers. During the generations we now realize were merely the dawn of the electrical age, the nuisance was confined to leakage in flashlights and "Batteries Not Included" with toys. Now, in terms of care and feeding, rechargeable batteries have raised the stakes. They rival pets in their demand for attention. At least the average Rottweiler owner can get by without much knowledge of inorganic chemistry. If you use Motorola's "two pocket automatic switching IntelliCharge II Rapid Charger" to speed the charging, you will need to keep your eye on the multicolor lamp in each pocket, bear in mind the difference between rapid charging and trickle charging, flip the little tabs on the batteries to track their status, and at all costs remember to remove batteries from the charger before twenty-four hours have passed. Specialty companies—battery boutiques—offer cheery tips: *Burp your battery. Clean your contacts. Work 'em out.* Perhaps without fully realizing it, you have taken on a new and complex management role: Strategic Coordinator of, and Footservant to, Batteries.

Total time spent? No one knows. But there is a possibility that your quotidian chores are expanding to fill a bit more than the available time.

Jog More, Read Less

Because you follow the guidelines of the President's Council on Physical Fitness and Sports, you spend a minimum of five to ten minutes a day warming up, ten to twelve minutes in slow stretching, and five to ten minutes cooling down—in addition to your two twenty-minute sessions each week lifting weights to improve muscular strength, three thirty-minute sessions of weight training for muscular endurance, and at least three twenty-minute bouts of aerobic activities—choose among brisk walking, jogging, swimming, cycling, rope-jumping, rowing, cross-country skiing, or games like racquetball and handball. That comes to at least forty-five minutes a day, spent merely to recreate the physical activity that, at least according to myth, came naturally in a healthier, primitive world. Here, too, you can *save* time, with the help of technology. Specialized machines promise to deliver concentrated workouts in just minutes. Brisk walking is, in itself, too slow. There is something out-of-kilter, anyway, about exercise as an organized use of time. As you head off for the eternal horizon on your treadmill, you must be aware that this march is almost by definition a waste of time, made possible by the luxury of time, made necessary by the disappearance of backbreaking labor from the daily routine. If you aren't puzzled by this paradox, perhaps you have never had to fight the weird impulse to fast-forward through the boring parts of the Jane Fonda video.

In more sedentary moments, you spend sixteen minutes a day reading books, and forty-one reading newspapers and magazines, by a typical gloomy estimate. That gets you through a small fraction of a book a week, and altogether much less text than in the past. Your grandparents might have read at least one newspaper in the morning and another in the evening. *USA Today* caters to your more modern reading habits by keeping its copy short. Other newspapers have catered to them by going out of business. *Slate,* an on-line magazine that began with an emphasis on deep political analysis, soon bent toward fast "service" material, meaning digests: "What if you could read five daily newspapers in under five minutes?" it advertised, promising essentially the same magic as an Evelyn Wood speed-reading tape. Even the *New York Times* altered its traditions to accommodate a time-pressured reading style, with modularized layouts and new tables of contents. This great purveyor of text conducted market surveys of people who matched *Times* readers socioeconomically in every way but one: they did not read the *Times.* Why not? The surveys found that the nonreaders felt intimidated by the sheer time-consuming bulk—that mass of words spread daily across broadsheet. Thus, without explicitly repudiating its "All the News That's Fit to Print" history, the newspaper began trying out a new slogan, still words of one syllable but fewer of them: "Read What You Like." Think of the newspaper as a sort of tasting menu. Get over the puritanical impulse to clean your plate.

You, of course, read more than the average. Even so, you catch yourself saying more and more, "I'm behind in my reading" . . . as though final term were ending and you had just discovered the mandatory reading list. The list is so long. More books on more subjects are published by more publishers every year. No topic is too small to be seen from several points of view. For example, the Internet bookstore Amazon.com featured, on the subject of the World Wide Web, more than nine hundred books by mid-1998—all, of course, quite new. It would be easy to say that publishers are simply churning out bad books, but it would be wrong. Impressive quantities of high-quality fiction pour into the bookstores, where they lean hard against equally impressive fiction published just months before and already wearing out their welcome on the shelves. The major chain bookstores return poorly selling titles to the publisher, as a matter of policy, after a period typically set at sixty days. Writers are wont to complain. "The shelf life of the average book is somewhere between milk and yogurt," Calvin Trillin has said. Every kind of retailer, for that matter, has learned to cater to shoppers with short attention spans and desire for stimulation. Management wisdom for the modern retailer includes offering "new, fresh product," keeping stores "exciting and innovative," and above all avoiding "the aging of inventory." A typical warning will read: "Excess quantities of old merchandise have impacted the stores' appeal and given customers the impression that there's 'nothing new.' " Yet the natural lifetime of a book, like the lifetime of a new disease wending its way dynamically through the population, depends on an incubation period that cannot be shortened. Books spread by word of mouth. If, on average, every reader of some book can give birth through persuasion to a mere 1.1 readers, that might create a readership: 1.1 raised to the 250th power is greater than the population of the world. But those generations take time, and for most books sixty days is not enough.

Recorded music has even less time to survive on the shelves, because of the efficient flow of information represented by Soundscan, a nationwide point-of-purchase tracking system. Soundscan has harmed the most creative side of rock and roll "because it has made information available too quickly to people who, in turn, overvalue the importance of information gotten too quickly," argues the music writer Gerry Marzorati. "Soundscan has killed off word of mouth. . . . It's wham-bam: An album isn't catching on in a week or two (as measured by Soundscan) and the plug gets pulled on promotion, the CD gets pulled from racks."

So many books and so many records compete for our limited attention. In 1947, when *Books in Print* began, it recorded a total of 85,000 books in print in the United States. By 1998, the number had risen twenty-fold, to 1.8 million, produced by no fewer than 44,000 different publishers. The American population, over the same period, had not quite doubled. No wonder a young online columnist confesses in a colloquy about J. D. Salinger that she could not make it through his bite-sized oeuvre: "Life is so short, and there's so much else to read, you know what I mean?"

Even reading to children is under pressure. Hence the 1983 volume *One-Minute Bedtime Stories,* traditional stories condensed, according to its publisher, "so they can be read by a busy parent in only one minute." You may feel that children themselves are not that busy. At least they may not need a stopwatch for the dash from once-upon-a-time to happily-ever-after, even if *Sesame Street* has done its work on their little psyches. But this book must have hit the mark, because many sequels followed, including *One-Minute Birthday Stories, One-Minute Teddy Bear Stories,* and *One-Minute Christmas Stories.* Perhaps the young targets of this bedtime largesse will grow up to join the ranks of those who consider a full-length book to be a quaint object. The columnist Russell Baker, watching a baseball game on television, heard the announcer ask a guest author whether she wasn't overwhelmed by the futility of writing a 300-page tome in an era when no one has time to read anymore. He wanted to shout: "Great Moloch, evil Baal and little fishes, man! Aren't you ashamed to ask a question like that in front of a huge audience that's about to spend three hours watching a televised baseball game?"

But—three hours? Only a senescent diehard fan wants a baseball game to last that long nowadays, or so Major League officials believe. Their desire to save time motivated almost every rule change of the 1990s. Batters were ordered to stay within three feet of the batter's box between pitches, to cut down on dawdling. Pitchers were ordered to throw within twelve seconds after the batter was set, down from the previous twenty seconds. Ballpark announcers and scoreboard operators were ordered to streamline their pre-inning routines. Organized baseball commissioned a time study by a former umpire, Steve Palermo; the next season, his recommendations cut an estimated six minutes from the average game time. When the strike zone was lowered through knee-level, it was in hope of forcing hitters to hack away instead of delicately prolonging those interminable full counts with foul ball after foul ball. The real cause of the lengthening baseball game was nowhere to be seen on the field: commercial breaks. But that did not stop aggrieved sportswriters from getting out their stop-watches and conducting efficiency-expert critiques of player behavior. Richard Sandomir described a typical Dodgers-Expos game (elapsed time, 3:05) as "a laboratory for observing how baseball's languid pace can turn to tedium." He timed a turn at the plate: eleven seconds after one foul ball, twelve more after another, fifteen more after another. He timed intervals between pitches, up to thirty-two seconds. He timed a fifty-two-second dugout-and-pine-tar break after a broken bat. He noted the "four to five seconds" it took a relief pitcher to spit and wipe sweat from his brow. He complained that Hideo Nomo, with his eye-twisting, hip-wrenching, double-stop delivery, took a full five seconds to pitch a ball across the plate—and complained that Nomo "wasted even more time" by preferring forkballs to fastballs. The sportswriter timed pitching changes—three in a single inning—of up to 120 seconds each. "One cannot help but wonder—once the two-minute pitching change is finally over—" he wrote, "if baseball's nineteenth-century tempo appears too anachronistic." Yes, it is anachronistic. Baseball is the only modern sport with more adagio than allegro built into the structure of the game. It is theoretically free of the clock—the only major spectator sport that can go on forever, now that tennis has instituted its own marathon-ending tiebreakers. Baseball has always included minor skirmishes of pace-breaking: batters withdrawing from their box, pitchers standing down from their mound, an Alphonse-and-Gaston war of wrecked tempos. Although balls and runners have moments of blurring speed, those are flashes in the dark, and a fan at the game can absorb a stadium-wide polyrhythm even when the ball itself is still: as Nomo clasps his hands together, wrenches his torso toward center field, winds the ball back over his head, and starts to lift his front foot from the ground, eight infielders and outfielders fill the passing microseconds with oblique maneuvers of their own—stepping sideways into gaps, ritually pounding gloves, tightening crouches, abandoning baserunners or sneaking behind them. It is characteristic of baseball that in a "perfect" game, no one gets anywhere. Tension replaces action, and the stopwatches may as well be put away.

No wonder baseball has lost its preeminent role in American culture. Perhaps it is not a game for modern times at that. Basketball, hockey, and the various footballs are faster. Anyway, watching sports has mostly become a subset of watching television, itself one of the biggest and roughest of time-use categories—"the 800-pound gorilla of free time," as one study puts it. In 1965, when Americans were averaging about an hour and a half a day, pollers asked whether they would like to watch more, or whether they would watch more if they could find better shows. Only 10 percent said yes. So a decade later, in 1975, the same pollers were shocked to discover that television time had risen by half, to more than two hours a day. Was it the arrival of color? Remote control? Cable? The fine quality of network programming?

No one could explain, but the trend continued steadily upward during the next decade as well. Estimates now surpass three hours a day.

Eat and Run

Time pressure has become the single most powerful force shaping the structure of the world's food industries: products, packaging, and marketing. More time goes to preparing food and cleaning up than to the actual eating of meals at home. Less and less time goes to all of these. People nibble in cars, at desks, and on street corners.

Packaged dry breakfast cereals, ready to douse with milk and eat, began as a fast alternative to cooked grains like oatmeal, instant or not. But pouring cereal into a bowl, adding milk, sitting down with a spoon—this process came to seem like foot-dragging. By the eighties, the breakfast-cereal market stagnated. Toaster food like frozen waffles and Pop-Tarts took over in some homes, before they, too, gave way to even faster food in the form of granola bars. The Kellogg Company tried to fight back on behalf of cereal with another form of innovation, Breakfast Mates, "for today's busy family," a single portion packaged with its own milk, bowl, and spoon. Dirk Johnson assembled his family one Sunday to test Breakfast Mates for the *New York Times* and found that they had managed to cut preparation time from fourteen seconds to thirteen, at a cost of one dollar per second saved.

If a science-fiction writer of the mid-century wanted to convey a bleak, sterile, inhuman future, a standard tactic was to describe a world whose subjects would consume all their essential daily nutrients in a few manufactured tablets—no dirty, irregular, bacteria-filled fruits, no vestigial tablecloths and silver candleholders. We are almost there, and it no longer strikes us as quite so bleak. We are consumers-on-the-run of brightly packaged superconcentrated protein drinks and foodstuffs: Powerfoods, Soy Delicious! Energy Bars, or Hammer Gel ("endurance fuels" featuring "protein powder"). Do you dare to eat a peach? The very notion of the family meal as a sit-down occasion is vanishing. Adults and children alike eat breakfast on the way to their next activity. Eggs, which once took minutes even to soft-boil, come in McMuffins or in frozen Toaster Scrambles. Even Pop-Tarts pop too slowly now. Dinner is not far behind on the road to obsolescence. Prepared, precooked, prepackaged meals—all the descendants of the TV Dinner—now take up more supermarket space than fresh fruits and vegetables. They threaten to surpass the rest of the traditional stock: the mere ingredients of meals. Teenagers lead the way, living lives as frenetic as their parents'. A seventeen-year-old says she eats relatively leisurely "homemade meals," meaning frozen pot pies and boxed macaroni-and-cheese, while her friends gobble takeout meals: "stuff they can eat right away or zap."

As the background pace of life in the kitchen has accelerated, other food products have gone from being time-savers to time-wasters: instant powdered lemonade was originally faster than squeezing lemons; now it is slower than opening bottles. Making cake frosting from a mix is faster than making it from scratch but slower than spooning it from a tin. Pancake and waffle mix saved only the time it took to add sugar and baking soda to flour, but that was enough—unless you prefer the further time-saving of frozen waffles and pancakes. Time-saving trajectories appear in the evolution of countless foods. Homemade frosting to frosting mixes to canned frosting. Gelatin-based desserts to Jell-O to premade Jell-O in jars. Packaged, frozen breakfast sausages to packaged, frozen, *precooked* breakfast sausages. With rising standards of living, the subtle tradeoffs of money and time have shifted in the direction of saving time. It is less expensive to ship premade soup in condensed form; but more and more consumers spare themselves the seconds it takes to add water or milk. Bouillon cubes came into the world as a leap forward in time-saving. Now, who can spare time to unwrap the foil and heat the water? For that matter, heating water, a process subject to constraints of physical law, has come to seem annoyingly slow. Hence the spread of Instant Hot and Quick & Hot faucets.

With a wealthier population of consumers and a rising tide of basic quality comes a war of attraction between quality and time-saving. Instant coffee, spooned from jars into boiling water, had its day, but the technology hit a wall, notwithstanding the space-age promise of "freeze-dried." In the realm of coffee, time-saving has advanced on parallel tracks. Go ahead and use real coffee. Percolators, once seeming fast, are now obsolete; the latest Brew 'n' CoffeeMaker cuts brewing and steeping time to a minimum. Or perhaps you are grabbing your high-quality brew, along with the rest of your breakfast, on the run at Starbucks. It will have a raised plastic top with a drinking hole, a "travel lid," instead of the flat top used on the quaint assumption that you planned to go someplace, sit down, open your coffee, and sip. Travel lids, like automobile cup-holders, are technology designed for the special purpose of drinking in motion.

Back in the kitchen, is your fresh pasta maker in active use? Your homemade ice cream machine? No, they are time sinks, so they have found a permanent spot on the rear shelf with the crock pot (*slow* cooking), while you rush to the supermarket for fresh-made tortellini and the latest Ben and Jerry's flavor. Or perhaps you just pick up the telephone. What began as an innovation of Chinese restaurants and pizza parlors—home delivery of hot food—has become high art. Takeout menus are the most pervasive form of door-to-door advertising in large American cities. Foods your grandparents never heard of arrive at your door in minutes, steaming in Styrofoam trays—tapas or rijsttaffel or hundred-dollar high-cuisine picnics. The delivery of pizza itself has become an international battleground. Fears of time-crazed drivers careering through the streets caused the Domino's chain to back away from a thirty-minute order-to-doorbell guarantee. When the Internet was young, one of the first popular services was a Pizza Server, constructing and transmitting pies from a thousand lines of C source code. It was only a minor drawback that the pies were virtual, rather than edible. At least they were fast. "Since the initial opening, the Pizza Server has only gone down once," the proprietors wrote proudly in 1994. "With the exception of that, and the patching of a small security hole, the Pizza Server has been running bug-free for nearly a year." Now anyone with a computer and modem can order real pizzas on-line, for delivery off-line, in Zurich, Madras, or Perth.

Total time spent eating? More for men than women, though the gap is closing. More for the unemployed. For all Americans, on average, just over an hour a day.

How Many Hours Do You Work?

It is work—the time-use category subject to the most diligent and official measurements—that finally breaks the back of any compilation of the typical day. Bureaucrats, economists, and academic sociologists are equally frustrated by the contrary messages from seemingly firm statistics. Here is American Studies—a typical college syllabus: "A generation ago Americans believed that their working hours would decline, their leisure increase, and their real incomes soar. As it has turned out, none of these expectations proved accurate. This course will examine why Americans today work more, shop for longer hours, and have less leisure time than they did in the 1960's."

A skeptical student will already be marveling that people who work longer for less money nonetheless *shop for longer hours.* But few have questioned the claim, widely repeated without qualification, that Americans are working harder and longer than ever before. "We have become a harried *working,* rather than leisure, class, as jobs take up an ever larger part of ever more Americans' lives," asserted the economist Juliet Schor in her 1991 book, *The Overworked American.* She calculated that the average employed American spent a full "extra month" working each year, compared with two decades earlier. This burden falls on men and, especially, women, on young and old, on full-time and part-time workers. People are moonlighting more, she said, and working longer at their main jobs. They are working more at home, cleaning their houses and caring for their children. They are taking less vacation and putting in more overtime. Adding insult to injury, they spend more time commuting, too.

It *feels* true, especially in certain high-profile professions. Lawyers saw their business transformed during these decades by the rise of the billable hour. Beginning in the early 1970s, firms made a transition to an intensely profit-driven hourly billing style, supported by new minicomputers with time-tracking software. They adopted "attorney productivity standards" based on hours billed. They remembered when one thousand hours billed had represented a respectable year's work, but by the end of the eighties, the lawyers at major firms in New York and Washington averaged more than eighteen hundred billable hours a year, and some firms achieved more. The real, audited number for Wachtell, Lipton, Rosen & Katz in 1983 was twenty-five hundred. That means fifty hours a week for fifty weeks, for every lawyer, not counting time spent in what one legal consultant called the dilly-dally zone:

> playing solitaire and minesweeper on your computer, reading newspapers and non-legal periodicals, calling family and friends who haven't called you in a while, calling into a radio talk show, coordinating activities for your nonprofit committee, talking about college football with your partner for forty minutes (bonus points for contributing to the ineffi-

ciency of the entire firm), going shopping for personal items, etc., etc., etc.

Not to mention eating, getting a haircut, playing a quick game of squash (or is the opponent a client?), going to the doctor, or just pausing to gaze unbillably out the window. In small firms, this consultant said, a lawyer who worked a ten-hour day would typically have trouble billing as many as five. No wonder bar association ethics panels warned against padding bills or working so many hours that fatigue might affect the work.

Lawyers were not alone in bringing the efficiency-expert mentality to bear on the allocation of every minute; nor in working long hours. Breakfast meetings at 8 A.M. entered some company routines; then the tough showed their mettle with 7 A.M. meetings. The new word *workaholic* implied a disease or a syndrome with a large involuntary component. Investment bankers notoriously sacrificed their waking hours to the deities of work. But university professors? Although an Ivy League faculty member who spent as much as ten hours a week (only thirty weeks a year) in actual classroom teaching would be heroic, professors at Penn State University, for example, reported working an average of fifty-two hours a week, and several claimed to work more than ninety hours. Perhaps for a professor thinking is work. And *workaholic* was the coinage not of a teacher or lawyer but of a minister, Wayne E. Oates, who noticed in 1968 that he and his colleagues were often compulsive, driven, restless, and positively addicted to their calling. God's work is never done.

It seems that some professions have developed a dynamic that defies the self-modulating textbook theories of wage equilibrium. According to the standard theories, unreasonable work weeks are self-defeating, even from the employer's point of view, because workers get tired or demand high overtime rates or simply rebel. However, according to a newer model, some businesses manage to develop what economists call a "rat-race equilibrium." The rat race occurs when managers use a willingness to work long hours as a sign of some intangible yet much-desired quality that merits promotion. An unstable negative feedback loop can arise. Employees who would really prefer to be catching the 5:15 train home for cocktail hour nonetheless try to disguise themselves as long-hour workers, at least for a while. Employers try to cut through the pretense by requiring even longer hours, even if that is inefficient. Employees who never leave have the right stuff. Managers reward not just the actual work product but the lights still on at night and the steaming coffee cup already visible on the desk at daybreak. After all, who feels the pressure more than the managers themselves? "Question," wrote the management guru Rosabeth Moss Kanter in the 1977 study *Men and Women of the Corporation*:

> How does the organization know managers are doing their jobs and that they are making the best possible decisions? Answer: Because they are spending every moment at it and thus working to the limits of human possibility.
>
> Question: When has a manager finished the job? Answer: Never. Or at least, hardly ever. There is always something more that could be done.

Every office an Augean stable. Conventional economic incentives do not apply. The rat-race, treadmill, fast-track players are trading every possible hour not for immediate wages but for a giant future reward, immeasurable in Invisible Hand terms: elite positions with profit shares, like partnership in a law firm, directorship of an investment bank, or tenure at a university. Dark Satanic mills!

Overwork at this far edge of the economy's spectrum of status and compensation may fail to inspire sympathy. Certainly the line between victim and perpetrator blurs. Every so often comes a dramatic dropout from the ranks, in dog-bites-man style—*financier forsakes Wall Street for bee-keeping*—showing, at least, that it can be done. These harried and harrying workaholics have chosen their poison. In New York and Los Angeles they consume it in the form of brilliantly abbreviated high-cuisine lunches: thirty minutes in million-dollar dining rooms that were conceived as settings for two- and three-hour *régalements*. The chefs swallow their pride, and the maître d's arrange a complex choreography for customers who pack in two lunches with successive guests. The resetting of tables for this bifurcated meal resembles the pit-stop activity at the Indianapolis 500. Then diners order their tuna rare and their potatoes microwaved to pare a few more minutes. Few will pity these casualties of hurry sickness. It's their own fault, for treating time as a mere status symbol. And a negative status symbol at that: the less time, the more prestige. The more time you have on your hands, the less important you must be. So sleep in the office. Never own up to an available lunch slot. The transformation of time into a negative status good has odd social consequences, as Michael Lewis has pointed out. "It boosts the credibility of things that happen quickly," he notes. "It also infuses with wonderful new prestige any new timesaving device. After all, who most needs such a device? People who have no time! And who has the least time? The best people!" And generosity has its limits. It's one thing to donate ten million dollars to charity. It's quite another to come to the phone ten seconds before a fellow executive whose secretary has placed the call.

Do we really believe that the rich and powerful spend all their time working? Of course not. "To work all the time is unpleasant; people rarely do it," says Lewis. "Instead, they fake it. Silicon Valley startup companies and Hollywood movies are especially useful for this purpose."

Even so, some economists believe that the attitude and the long hours, whether real or faked, have trickled down through the workforce, to those who dine in the company cafeteria or the neighborhood hot dog vendor or who serve as their own maître d's at their desks, obliterating any trend that might have led to a society of leisure. A harsh business system is forcing the vast majority of workers, salaried and unsalaried, full-time and part-time, to extend their hours, or so these economists argue. They line up many kinds of statistics in support of this conclusion. The most official come from the federal government: the Census Bureau's Current Population Survey samples tens of thousands of Americans monthly, asking them to estimate their work hours, and the Bureau of Labor Statistics gets payroll numbers from businesses. In their raw form, these data tend not to show a rising trend. On the contrary, payroll records show a steady decline in weekly hours worked over the four decades they have been kept, and even the subjective responses of workers about their own time have been almost flat—down, overall, about two hours since the 1950s. Still, these are averages, and they are flawed in so many ways that economists will adjust them and cavil about them forever. Changing patterns of work, and changing definitions, make any indisputable analysis impossible. More people are working at home, more people are working part-time, more people are working two jobs, more people are working as independent contractors, and all these trends add to the statistical morass.

How many hours did you work last week? The Census Bureau may ask you.

Do you count commuting time? Lunch time? Breaks? If you work a traditional nine-to-five shift, statistics suggest that you will call that a forty-hour week—so perhaps you count every minute on the job as work, whether you take a lunch break or not. If you are on an assembly line in Dearborn, Michigan, your time is closely monitored; if your work moves from meeting to memorandum to telephone call, it is just possible that your desktop computer has found room for a game, with an instant hotkey to change the screen if your boss appears behind your back. Either way, if you missed some scheduled work time last week—a doctor's appointment, a household emergency, a sick day, a holiday, a broken-down car—do you subtract that time, or do you give the Census Bureau a more official number?

The Census Bureau's methodology avoids complexity:

> How many hours did you work last week?
> Your answer: ____ hours.

Luckily, the form does not ask how many hours you spent playing computer games. But no matter how simple and consistent the question, collecting real data about how people use their time is a procedure riddled with difficulties. Time-use statistics that come from interviews rely on people's memories and their ability to estimate large and small chunks of time. Alternatively, they can come from actual observation or telephone sampling, but these approaches do not help the Current Population Survey. Statistics can also come from clever computations: total airline passenger miles divided by average speed, or total payroll hours divided by the worker population.

The most comprehensive academic research on how people spend their time comes from a decades-long historical project at the University of Maryland called the Americans' Use of Time Project. The project's leaders, John P. Robinson and Geoffrey Godbey, believe they have found silly and grotesque flaws in the other methods of calculating time use, especially where work is concerned. "People think they know how many hours they work," the professors say, "—that is, until they actually try to figure it out." It is one thing to give a snap answer to a telephone interviewer. Submitting to the cross-examination of the Use of Time Project is different. Subjects sound like this:

> So, not including that one hour off and the nine hours off I think I worked, like, forty-one and a half, including that time off. So, minus nine is—thirty-two. I think I worked

> thirty-two and a half—something like that. OK. Oh, god . . . my schedule goes from Thursday to Wednesday . . . Let me do it backwards. Did I work Saturday? Yes, I worked Saturday. Sunday to Saturday or Saturday to Sunday? Sunday to Saturday. Saturday I worked from 6 to 10, and I worked Friday—no, Thursday—yes, I worked 12 to 4:30. Wednesday—yes, I worked—when did I work? Nah, I volunteer-worked that night. No, I didn't work. Sunday, did I work? Oh gosh, Sunday night . . . gosh, did I work that day? I think I may have worked that day. What did I do? I watched the football game? . . . So that's four, four and a half, and six, ten and a half—I'll say fourteen and a half hours.

An accurate calculation? Who knows? Yet the same person, answering the Census Bureau's quick question, might have given the short answer, forty-one—a lot more than fourteen.

We truly believe we are busy. We know we are busy. But if the time-use research demonstrates anything for certain, it demonstrates that we don't know *how* busy, in terms of minutes, hours, or days. In the quasi-reality of this research, how much of the twenty-four-hour pie chart goes to work, then? A bit under six, for employed men and women, according to Robinson and Godbey. Quite a bit less if the nonworking population is included for the sake of an average. They see a clear downward trend in hours spent working, from 1965 to 1975 to 1985, and many economists had already come to the same conclusion through different paths. Juliet Schor disputes their numbers, and the government, for all its gathering of labor statistics, just doesn't know. Schor says, "My estimates, which are both more comprehensive and representative of the U.S. population than those of previous researchers, reveal a clear and dramatic trend to more work." On that time scale, Schor sees "the growth of work," "the work explosion," and "the extra month of work." Many reviewers and readers believed her; they wanted to believe her. She strikes a sympathetic note. When people are asked to make their own estimates, they do think their work fills more time than in the past—an undated and probably mythical past. Not only that, but the harassed laborers boast about their busy-ness. Overwork equals importance. An overfull schedule is a talisman of status and rank. When two people negotiate a lunch date, they must each be careful not to concede a more open calendar. A peculiarly modern righteousness comes with that fifty-hour—no, sixty-hour—no, seventy-hour work week.

In reality, the claims of a work explosion are unsupportable. There was no segment of the American work force of the sixties and seventies that had an "extra month" (whether Schor means one-twelfth more or one-eleventh more hours) free to fill in with more work. Where did that monumental slab of new work time come from? Has leisure given way?

Not television watching. Not time on the StairMaster. Not time spent driving. Not time spent figuring out how to program the VCR (or watching it; video didn't exist—remember?). Not time spent playing computer games—in 1998, the average personal-computer user put in an estimated 10.3 hours a month at a single game, Civilization II; then there is Myst or Riven or Doom or Quake (anyone who has experienced these time sinks, nonexistent in 1970, knows that they can make hours flash by, gripping the mind with an addiction more powerful than any strain of workaholism). Not time at the National Parks—although the average hours spent per visitor has indeed dropped slightly over the last generation, and tourists do seem to race along those trails, many more consumers of nature have been able to travel to the parks, a twelve-fold increase over four decades; they fill well over a billion hours a year. Not time spent gambling—a third of American households now visit a casino each year, and they invest more than money: each trip means, on average, eighteen hours per person at the slots or the tables. If leisure means free time—truly free; free of Myst and Quake, free of hiking and reading and listening to music—then perhaps we have lost our dream of leisure. We do have time, free or not, that we like to fill with recreation.

Yet Schor says that perhaps "work itself has been eroding the ability to benefit from leisure time." We're just too tired to relax, too tired to have fun. It's no coincidence, she says, that the most popular ways to spend the evening are "low-energy choices," meaning television, which she implicitly scorns. In this way, the economists whose statistics suggest a work explosion tend to view workers as victims—victims of something big and systemic and inhuman. Some of us are perpetrators, too. All in all, we are a mixture of victims and perpetrators, with nothing in between. Women are victims almost by definition. But even men are victims. "Men are ensnared," Schor writes—victimized by a "role" and also by a "tendency of our culture." For the perpetrators—such as people who refuse to trade some of their wages for extra free time—she has harsh language. They are "workaholics," or those "for whom money is everything," and they "sell their souls to the highest-paying jobs they can find." Then again, maybe they are making a free and deliberate choice.

7:15. Took Shower

The day was already full. However long you work, by the time you have totaled the figures for sleeping, shaving, dressing, driving, talking on the phone, reading, eating, exercising, searching for lost objects, waiting for the computer to boot, glancing at your family, and watching television, you have far surpassed twenty-four hours. It's impossible. And what if you want to kill a few minutes by listening to an old record? What if you need to run over to the post office and buy some stamps? What about prayer (yes, there are statistics for this)? If you don't bend your knees, what about unmonitored secular contemplation (no statistics, but we do still daydream)?

Government agencies, think tanks, company researchers, and academic sociologists all pursue the mission of creating statistics on how people use time. Any one example tends to be convincing enough. As the statistics accumulate, however, they begin to appear contradictory, self-serving, meaningless, and wrong. Just as the average American will die several times over if one adds the advertised risk percentages associated with the most prominent diseases and accident types, so the time-use pie chart inevitably bursts through the twenty-four-hour mark and keeps on overflowing. Time uses, like diseases, have passionate constituencies. Caveat emptor—and beware, too, of the pie chart itself. Tempting though it is to think of the average day as an object with a certain bulk, ready to be sliced and divided among hungry competing activities, time just does not work that way. The pie chart automatically distorts reality.

The easiest way to find out how much time people spend on an activity is to ask them. If only they could remember! One of the flaws with this technique is that people prefer to think about a typical day—a day that never comes. Despite standard time, despite synchronicity, despite the nine-to-five day and the 11:30 P.M. talk shows, we weave our own twisted pathways in and out of the available minutes. The two questions "How much time *do* you spend reading *each day*?" and "How much time *did* you spend reading *yesterday*?" produce sharply different results in surveys. For reading and most other activities, *each day* is the unreal typical day and *yesterday* usually comes up short. Maybe you feel you spend a half-hour a day reading, but yesterday for some reason you were never able to pick up that book. There is always some reason. No matter how honest responders try to be, the idealized day overshadows the real one. So researchers try other techniques. They hire assistants to follow their subjects around anthropologically—labor-intensive, and useful only with the sort of subjects who have time to be followed around. Or the researchers attach electronic tags to the subjects, to verify whether they are, say, in range of the television set. Or researchers dial the telephone and ask subjects what they are doing—that is, what they were doing before they answered the phone. The telephone spot-check is popular, but it inevitably undercounts activities that keep people from answering their phones. No one ever says they are out hiking in the woods, and few admit to having interrupted an act of sex.

People have poor time sense during many of these activities. But at least in the case of television, subjects have a handy measuring stick right there. They know how long a program lasts. When calculating television time, they can think, *Seinfeld*—OK, a half-hour. This handy shortcut does not work for *Finnegans Wake.* Even so, Nielsen raters have found the measuring of television time an endlessly deep problem, demanding the application of ever more technology: the notebooks, set-top boxes, networked set-top boxes. None of these devices can tell whether the subject closed the shades and sat down in the easy chair to concentrate on the final episode of *Roots,* or whether the television set is just on, as in so many households, like a noisy light bulb, while the life of the family passes back and forth in its shimmering glow? Perhaps it would be best if researchers could wire every viewer's prefrontal lobe directly to the Nielsen operations center in Dunedin, Florida.

The Americans' Use of Time Project has used data from many sources to monitor trends over the final third of the twentieth century, but has relied most carefully on time diaries. These are journals kept by thousands of subjects, who log their activities minute by minute, in something approaching real time. The project's experts, Robinson and Godbey, call their technique a "social microscope." The diaries have flaws of their own. Few participants have the patience to list more than thirty or forty different activities, so the picture lacks fine granularity. Typical entries are:

5:45 AM–6:00 AM Did stretching exercise.

. . .

6:20 AM–7:00 AM Went to health club, exercising.

. . .

7:15 AM–7:35 AM Took shower.

These are still broad segments on the surface of the subjects' lives. If only it were feasible to drill down further to the entries that *don't* exist:

1:22:00 PM–1:22:48 PM	Carried leftovers to microwave.
1:22:48 PM–1:23:27 PM	Answered phone. Begged telemarketer to remove me from list.
1:23:27 PM–1:24:27 PM	Warmed food for 60 seconds. Grabbed newspaper and read headlines, meanwhile.
1:24:27 PM–1:24:56 PM	Ate. Recycled newspaper. Planned next time-diary entry.

There would be no endpoint. The process would reveal itself as fractal and recursive. Let's look more closely at the impossible, hypothetical 1:23:

1:23:34 PM–1:23:38 PM	Thought about work.
1:23:38 PM–1:23:40 PM	Daydreamed.

At this level of detail, daydreaming may become a more and more persistent activity in any species of conscious being. The atoms of time use can never be found.

Then, the researchers must distribute the entries among general categories. Food preparation. Reading. Housework. Personal care. Free time. They are often forced to confront the arbitrary nature of their decisions. "Human behavior is potentially infinite in meaning and form," Robinson and Godbey admit in a philosophical moment. "What are you doing is, ultimately, an existential question." What are you doing? How much time are you spending? The act of measurement can lead to obsession. No wonder that as the researchers look around, they see a rushing and scurrying everywhere: "Sometimes American culture resembles one big stomped anthill." In a kind of anti-Zen parable, they report that one of their own colleagues stopped to calculate the time he spent tying shoes and buckling belts. He projected this number out through the rest of his presumed life span and, horrified, made a decision to cut back. Henceforth, they say, he has worn only Sansabelt pants and Velcro'd sneakers. Call it *rushwear.*

At least the time diaries are a zero-sum game. They leave room for a reliable 1,440 minutes a day. The first entry begins at midnight; the last entry ends at midnight. Pass GO. Collect twenty-four hours.

Attention! Multitaskers

The final, fatal flaw in the time-use pie chart is that we are multitasking creatures. It is possible, after all, to tie shoes and watch television, to eat and read, to shave and talk with the children. These days it is possible to drive, eat, listen to a book, and talk on the phone, all at once, if you dare. No segment of time—not a day, not a second—can really be a zero-sum game.

"Attention! Multitaskers," says an advertisement for an AT&T wireless telephone service. "Demo all these exciting features"—namely E-mail, voice telephone, and pocket organizer. Pay attention if you can. We have always multitasked—inability to walk and chew gum is a time-honored cause for derision—but never so intensely or so self-consciously as now. If haste is the gas pedal, multitasking is overdrive. We are multitasking connoisseurs—experts in crowding, pressing, packing, and overlapping distinct activities in our all-too-finite moments. Some reports from the front lines:

David Feldman, in New York, schedules his tooth-flossing to coincide with his regular browsing of on-line discussion groups (the latest in food, the latest in Brian Wilson). He has learned to hit Page Down with his pinky. Mark Maxham of California admits to even more embarrassing arrangements of tasks. "I find myself doing strange little optimizations," he says, "like life is a set of computer code and I'm a compiler." Similarly, by the time Michael Hartl heads for the bathroom in his California Institute of Technology digs each morning, he has already got his computer starting its progress through the Windows boot sequence, and then, as he runs to breakfast, he hits Control-Shift-D to dial into the campus computer network, and then he gets his Web browser started, downloading graphics, so he can check the news while he eats. "I figure I save at least two or three minutes a day this way," he says. "Don't laugh." Then there's the subroutine he thinks of as "the mouthwash gambit," where he swigs a mouthful on one pass by the sink, swishes it around in his mouth as he gets his bicycle, and spits out as he heads back in the other direction, toward a class in general relativity.

The word *multitasking* came from computer scientists of the 1960s. They arranged to let a single computer serve multiple users on a network. When a computer multitasks, it usually just alternates tasks, but on the finest of time scales. It slices time and interleaves its tasks. Unless, that is, it has more than one processor running, in which case multitasking can be truly parallel processing. Either way, society grabbed the term as fast as it did *Type A*. We apply it to our own flesh-and-blood CPU's. Not only do we multitask, but, with computers as our guides, we multitask self-consciously.

Multitasking begins in the service of efficiency. Working at a computer terminal in the London newsroom of Bloomberg

News, Douglas McGill carried on a long telephone conversation with a colleague in New York. His moment of realization came when, still talking on the phone, he sent off an E-mail message to another colleague in Connecticut and immediately received her reply. "It squeezes more information than was previously squeezable into a given amount of time," he says. "I wonder if this contributes to that speeding-up sensation we all feel?" Clearly it does.

Is there any limit? A few people claim to be able to listen to two different pieces of music at once. Many more learn to take advantage of the brain's apparent ability to process spoken and written text in separate channels. Mike Holderness, in London, watches television with closed captioning so that he can keep the sound off and listen to the unrelated music of his choice. Or he writes several letters at once—"in the sense that I have processes open and waiting." None of this is enough for a cerebral cortex conditioned to the pace of life on-line, he realizes:

> Ten years ago, I was delighted and enthralled that I could get a telegram-like E-mail from Philadelphia to London in only fifteen minutes. Three years ago, I was delighted and enthralled that I could fetch an entire thesis from Texas to London in only five minutes. Now, I drum my fingers on the desk when a hundred-kilobyte file takes more than twenty seconds to arrive . . . damn, it's coming from New Zealand . . .

It seems natural to recoil from this simultaneous fragmentation and overloading of human attention. How well can people really accomplish their multitasks? "It's hard to get around the forebrain bottleneck," said Earl Hunt, a professor of psychology and computer science at the University of Washington. "Our brains function the same way the Cro-Magnon brains did, so technology isn't going to change that." But for many—humans, not computers—

a sense of satisfaction and well-being comes with this saturation of parallel pathways in the brain. We divide ourselves into parts, perhaps, each receiving sensations, sending messages, or manipulating the environment in some way. We train ourselves as Samuel Renshaw would have trained us. Or, then again, we slice time just as a computer does, feeding each task a bit of our attention in turn. Perhaps the young have an advantage because of the cultural conditioning they received from early exposure to computers and fast entertainment media. Corporate managers think so. Marc Prensky, a Bankers Trust vice president, had to learn to overcome instinctive annoyance when a young subordinate began reading E-mail during a face-to-face conversation; the subordinate explained: "I'm still listening; I'm parallel processing." This whole generation of workers, Prensky decided, weaned on video games, operates at *twitch speed*—"your thumbs going a million miles a minute," and a good thing, if managers can take advantage of it.

At least one computer manufacturer, Gateway, applies multitasking to technical support. Customers call in for help, wait on hold, and then hear voices. "Hello," they are told. "You are on a conference call." William Slaughter, a lawyer calling from Philadelphia, slowly realizes that he has joined a tech-support group therapy session. He listens to Brian helping Vince. Next, Vince listens to Brian helping William. It's like a chess master playing a simultaneous exhibition, William thinks, though Brian seems a bit frazzled. Somehow the callers cope with their resentment at not being deemed worthy of Brian's undivided attention. Why should he sit daydreaming while they scurry to reboot? "Hello, Vicky," they hear him say. "You are on a conference call."

There is ample evidence that many of us choose this style of living. We're willing to pay for the privilege. An entire class of technologies is dedicated to the furthering of multitasking. Waterproof shower radios and, now, telephones. Car phones, of course. Objects as innocent-seeming as trays for magazines on exercise machines are tools for multitasking (and surely television sets are playing in the foreground, too). Picture-in-picture display on your television set. (Gregory Stevens, in Massachusetts: "PIP allows me to watch PBS/C-Span or the like, and keep the ball game on or an old movie. Of course, it is impossible for anyone else to enjoy this, with me changing the pictures and audio feed every few seconds. When the computer and the phone are available in a multi-window form on the television, things are going to be very different.") Even without picture-in-picture, the remote control enables a time-slicing variation on the same theme. Marc Weidenbaum, in San Francisco, has a shorthand for describing an evening's activities to his girlfriend: "Got home. Ate some soup. Watched twenty or thirty shows." He means this more or less literally:

> I'll watch two sitcoms and a *Star Trek: Voyager* episode and routinely check MTV (didn't they used to run music videos?) and CNN (didn't they used to run news?) in a single hour.
>
> And really not feel like I'm missing out on anything.

Nothing could be more revealing of the transformation of human sensibility over the past century than this widespread unwillingness to settle for soaking up, in single-task fashion, the dynamic flow of sound and picture coming from a television screen. Is any one channel, in itself, monotonous? Marshall McLuhan failed to predict this: the medium of television seemed *cool* and all-absorbing to him, so different from the experience available to us a generation later. For the McLuhan who announced that the medium was the message, television was a black-and-white, unitary stream. McLuhan did not surf with remote control. Sets were tiny and the resolution poor—"visually low in data," he wrote in 1964, "a ceaselessly forming contour of things limned by the scanning finger." People were seen mostly in close-up, perforce. Thus he asserted: "TV will not work as background. It engages you. You have to be *with* it."

No longer. Paradoxically, perhaps, as television has gained in vividness and clarity, it has lost its command of our foreground. For some people television has been bumped off its pedestal by the cool, fast, fluid, indigenously multitasking activity of browsing the Internet. Thus anyone—say, Steven Leibel of California—can counter McLuhan definitively (typing in one window while reading a World Wide Web page in another): "The Web and TV complement each other perfectly. TV doesn't require much attention from the viewer. It fits perfectly into the spaces created by downloading Web pages." If he really needs to concentrate, he turns down the sound momentarily. Not everyone bothers concentrating. Eight million American households report television sets and personal computers running, together in the same room, "often" or "always."

Not long ago, listening to the simpler audio stream of broadcast radio was a single-task activity for most people. The radio reached into homes and grabbed listeners by the lapel. It could dominate their time and attention—for a few decades. "A child might sit," Robinson and Godbey recall sentimentally, "staring through the window at the darkening trees, hearing only the Lone Ranger's voice and the hooves of horses in the canyon." Now it is rare for a person to listen to the radio *and do nothing else.* Programmers structure radio's content with the knowledge that they can count on only a portion of the listener's attention, and only for intermittent intervals. And rarely with full attention. Much of the radio audience at any given moment has its senses locked up in a more demanding activity—probably driving. Or showering, or cooking, or jogging. Radio has become a secondary task in a multitasking world.

Shot-Shot-Shot-Shot

All the media have felt the acceleration. Hot media, cool media—it no longer matters. You visit the production set of a Hollywood movie on its last day of shooting. Cameras roll, an elevated platform shakes, and Sharon Stone—her famous face all but obscured inside a black diving helmet—widens her eyes in ersatz fear. She is sitting with Dustin Hoffman to her left and Samuel L. Jackson to her right, all crushed inside a flimsy plastic bubble raised high off the floor of a Warner Brothers production set in an old Navy base in northern California. Her director, Barry Levinson, is speaking to her in real time through a hidden earpiece. He is one of America's most thoughtful and word-oriented directors; he has recently finished filming *Wag the Dog* at a record pace, in a matter of weeks, with a waggish screenplay by David Mamet lampooning Washington and Hollywood together with fast streams of dialogue. But now Levinson is saying: "Fireball is coming up, coming up . . . coming straight toward you . . . wham! In your face!" And Stone grimaces appropriately as a beam of yellow light flares from somewhere beneath her right foot. One shot completed.

When the movie, *Sphere,* is finished, and the green-screen background is digitally replaced with computer-processed rushing water, the bubble will pass convincingly for a miniature submarine. A submarine in a hurry, you can tell: on a panel behind the actors, radiant red numerals flash the passing time in tenths of a second. (Tenths are also the new standard for that when-all-else-fails mechanism of suspense, the bomb with its own handy clock display. Some films, presumably with tongue in cheek, flash their end-of-the-world countdowns in hundredths of a second, just a blur of numerals.)

The crew sets up again for the shot. "We'll go tumble, tumble, tumble, tumble; then suddenly here goes the fireball," says Levinson. An assistant, counting, explains, "That's four tumbles." One of the two active cameras, the Hot Head Plus Dutch, is mounted on gimbals so it can spin through 360 degrees and on rails so it can rush toward the minisub, creating the illusion of motion at implausibly high speed. Only the most cynical of viewers will consider that a submarine so bulbous and unstreamlined could never cut through water this fast. "Stand by for the shaking—hold on, everybody," the director shouts, and the actors brace their gloved hands and black boots inside the bubble. They speak their lines all together: "We're going into it!" "Pull up, pull up, pull up!"

Levinson watches through both cameras at once via remote television monitors—on the modern film set, there is no waiting around for "dailies." It's still not quite right. "Can we spin faster?" he says. "Spin faster!"

No matter how fast a movie goes these days—or a situation comedy, a newscast, a music video, or a television commercial—it is not fast enough. Vehicles race, plunge, and fly faster; cameras pan and shake faster, and scenes cut faster from one shot to the next. Some people don't like this. "Shot-shot-shot-shot, because television has accustomed us to a faster pace," says Annette Insdorf, a Columbia University film historian. "There's a kind of mindlessness. The viewer is invited to absorb images without digesting them. Music videos seem to have seeped into the rhythms of creativity. It's rare these days that films afford the luxury of time."

Television, too, is behaving like a horse with a methamphetaminic rider. A new forward-looking unit within NBC, called NBC 2000, has been taking an electronic scalpel to the barely perceptible instants when a show fades to black and then rematerializes as a commercial. Over the course of a night, this can save the network as much as fifteen precious seconds, even twenty, but that is not the real point. The point is that the viewer, at every instant, is in a hurry. That's *you* pressing the gas pedal. Give you a full second of blank screen, and your thumb starts to squeeze the change-channel button . . .

New technologies, in living rooms and in editing studios, are helping to drive the pace of art and entertainment, just as they are driving the pace of virtually everything else in our work lives and our leisure time. Levinson is not a director of action movies—on the contrary, his best work (*Diner, Rain Man, Avalon*) has flowed at the pace of human character growth, on distinctly nondigital backgrounds with rich emotional texture. In these films, the clocks didn't need second hands. But here he is, in a darkened hangar, shooting the purest action sequence of his career, eyes on the monitors as three fine actors hurl themselves from side to side in the style of the troupers on the bridge of the starship *Enterprise*. We will not linger. We will

CUT TO:

INT. THE DIRECTOR'S TRAILER—DAY

where the same Barry Levinson is lamenting the summer of *Speed 2*—and for that matter the whole notion of bang-zip-pow "summer movies." Do our brains stop working in summer? "It's not an accident that all the movies of the summer are *rides,*" he says. "Adrenaline! Our rhythms are radically different. We're constantly accelerating the visual to keep the viewer in his seat." The restless viewer is very much on the filmmakers' minds—though at least in the movie theater they can expect viewers to stay in place for the allotted hundred minutes.

"I don't know that we demand more content—we demand more movement," says Levinson. "We're packing more in, but the irony is that it isn't more *substance.* We all become part of that. We all become less patient."

And . . . why? Well, there is television. "You cannot put a child in front of a television set where he is bombarded by images and not ultimately have an adult who is born and bred to see things differently," he says. "How can that not alter us?"

To older critics, who grew up with what now seems a methodical and plodding style of film storytelling, it seems as if we are engaged in a vast psychology experiment conceived by a sadistic professor who assaults the subjects with visual images at a rate up to and beyond the limits of perception. A generation ago, the word *subliminal* came into vogue, as in "subliminal advertising," with a fear that images could flash by so fast that we might see them, and come under their sway, without quite *seeing* them. Now we're used to subliminal imagery. We don't get scared when a commercial for Nike or Pepsi goes off on our screen like a string of firecrackers, but still, how much do we comprehend? How do we feel afterward? What will we want next? Reviewers talk routinely now about visual candy and visual popcorn, of the sinews of plot and character melting away in a boil of visceral gratification. In 1982 Pauline Kael, in *The New Yorker*, assailed the turn to hyperactivity represented, for her, by *Star Wars* and *Raiders of the Lost Ark*—two films that stretched the limits for fast-action sequences. At the century's end, these films already seem like classics; they had structure, characters, and wit. Now we have what

Anthony Lane, one of Kael's successors, calls "our own ever-growing predicament: there is nothing so boring in life, let alone in cinema, as the boredom of being excited all the time."

Levinson's own television series, *Homicide: Life on the Streets*, got the attention of critics with its frenetic, jittery camera style—the Point of View leaping about so assertively that Levinson sometimes had trouble getting his own editors to cut the scenes the way he wanted. Not only does the camera jump around within a scene, but shots are quickly interleaved from different angles to show multiple split-second views of the same scene—"double cutting" or "triple cutting." This style deliberately interrupts the continuity that filmmaking long took as a goal. Continuity meant realism—the illusion of life passing at normal speed. In *Homicide* the fast cutting and unstable cameras are meant to convey a different, gritty, true-life realism. But the style can be completely divorced from any particular tone or mood. Dramas and situation comedies alike try to ensure that their characters are on the move as they exchange the next bit of dialogue—striding, or better, running, along the sidewalk or down the office corridor. No talk without simultaneous action, because we viewers, after all, enjoy multitasking, and we can absorb the words and movement all at once. Watch new-format chat shows like *Loveline* on MTV and you see the same uneasy visual style, cameras constantly on the go, even tilting from side to side. (Have they run out of tripods? Are we supposed to wonder about the camera operators' sobriety?) The content is nothing but people talking, yet no shot seems to last more than a second. If it did—if the camera actually settled on one person's face for the time it takes to speak a full sentence—would you change the channel? The programmers think so.

We have learned a visual language made up of images and movements instead of words and syllables. It has its own grammar, abbreviations, clichés, lies, puns, and famous quotations. Masters of this language are the artists and technicians, Muybridge descendants, who create trailers for movies and thirty-second commercials and promotional montages of film clippings. And we in their audiences are masters, too, understanding the most convoluted syntax at a speed that would formerly have been blinding. What we see, we use to just *see*, the light streaming in through the eyes in real time only. Now we manipulate it, break it up, rerun it, and, of course, accelerate it. We absorb information in volume, with true virtuosity. This language continues to evolve. We see the situation-comedy hero sitting smugly behind his desk. He is told, "No, she's got *your* job." Instantly we see the same office stripped to a bare desk with the heroine neatly brushing away some dust—and that joke was quicker to see than to read. Not everything is faster; on the soundtrack, few modern films feature the kind of exhilarating machine-gun dialogue that filled the screwball comedies of the thirties, by Howard Hawks and others. Then, the camera's eye had to remain more or less fixed (heavy cameras) while Cary Grant and Rosalind Russell volleyed repartee across the screen. The new technology of radio had forced briskness and brevity on professional speakers, such as politicians, who were accustomed to orating on the stump for three hours at a stretch, and preachers, sometimes drilling words into their listeners at speeds that reached two hundred words a minute. But even the rat-a-tat-tat of Walter Winchell on the radio and the breakneck wordplay of Groucho Marx ("you can leave in a taxi—if you can't leave in a taxi you can leave in a huff—if that's too soon you can leave in a minute and a huff") lags when compared with any modern comic monologue of Robin Williams—jokes, allusions, whole personas flying past the ear at nearly subliminal pace.

Psychologists note that, while a normal fast-talker speaks at up to 150 words a minute, listeners can process speech reaching the ear at 500 or 600 words a minute, three to four times faster. *Can* and, these days, *want to*. That's a big gap, between how fast people talk and how fast people hear. This gap accounts for auctioneers and race-track announcers and now for the fast-playback button on telephone answering machines and for the fast-talking shtick of John Moshitta, who reached his summit of popularity, appropriately enough, in famous Federal Express commercials. The gap also creates an opening where ennui creeps in. A normal human being speaking at normal speed—the President of the United States, say, taking a full hour to deliver the State of the Union message—is less likely than ever to deliver the constant punch needed to hold our attention. History, entertainment—it hardly matters. Within hours of the opening of the only presidential impeachment trial in the twentieth century, the *Washington Post* assessed its pace with the page-one headline, "On the Floor, the First Day Wore On, On, On," and a comment by its television critic: "Who'd have thought that making history could be such excruciatingly ponderous torture?" People talk too slowly. Our minds race on like runaway conveyor belts past hapless Lucy Ricardos struggling with the chocolates. Fill us up, faster! Mere conversation, in front of an inert camera, doesn't seem to do the trick.

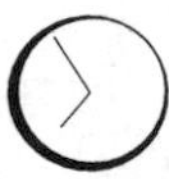

Prest-o! Change-o!

We have acquired various hand-held antiboredom devices: chiefly, the "remote." Television watchers jump from channel to channel, and filmmakers copy that by jumping from scene to scene. The more we jump, the more we get—if not more quality, then at least more variety. Saul Bellow, naming our mental condition "an unbearable state of distraction," decided the remote control was a principal villain.

> Pointless but intense excitement holds us, a stimulant powerful but short-lived. Remote control switches permit us to jump back and forth, mix up beginnings, middles and ends. Nothing happens in any sort of order. . . . Distraction catches us all in the end and makes mental mincemeat of us.

When the first remote controls appeared in the 1950s, as luxury add-ons for television sets, they seemed like innocent devices that would save viewers occasional trips from the bed or sofa to the television set. They were pitched at the lazy or infirm. "Prest-o! Change-o! Remote control tuning with 'Lazy Bones' station selector," said a Zenith advertisement. "Amazing!" The inventors and marketers thought the primary purpose of their device would be to turn the set *off* as the user drifted toward sleep. Secondary uses, they thought, could include silencing commercials and, sure, changing channels, presumably once or twice an evening, when programs ended. (*Consumer Reports,* comparing the first models, sniffed that the magazine "did not test, though it recommends judicious use of, a simple built-in control device present on every television and radio set known as the 'off-switch.' ") Marketers tried not-so-subtle appeals to masculine gun and control fetishism—users could "zap" with the "Flash-Gun" and "Space-Commander." No one imagined the real power waiting in the remote control. The advertising and commentary of the fifties shows that it was not seen as a time-saving device in any sense. Nor did anyone think in terms of amplifying the television experience with dozens or thousands of channel changes per evening. Most households could get just three to five channels; how could they imagine the remotemeisters waiting a generation up the road, using their wands to create on-the-go montages, nightly sound-and-light shows?

Now every television programmer works in the shadow of the awareness that the audience is armed. The remote control serves as an instant polling device, continually measuring dissatisfaction or flagging attention, if not for Nielsen's benefit then for your own. Possession of the device means that you have a choice to make every second. *Is this dull? Am I bored yet?*

The remote control is a classic case of technology that exacerbates the problem it is meant to solve. As the historian of technology Edward Tenner puts it: "The ease of switching channels by remote control has promoted a more rapid and disorienting set of images to hold the viewer, which in turn is leading to less satisfaction with programs as a whole, which of course promotes more rapid channel-surfing." If only the programmers could tie your hands . . . for your own good! Still, isn't possession of the remote a form of power? It does serve you, as a weapon against bad programming, even if the audience does not always use it wisely. Robert Levine, a social psychologist, cites studies that find "grazers" changing channels twenty-two times a minute. "They approach the airwaves as a vast smorgasbord, all of which must be sampled, no matter how meager the helpings," Levine writes. He contrasts these frenetically greedy Westerners—Americans, mostly—with Indonesians "whose main entertainment consists of watching the same few plays and dances, month after month, year after year," and with Nepalese Sherpas who eat the same meals of potatoes and tea through their entire lives. The Indonesians and Sherpas are perfectly satisfied, Levine says.

Are they really? Will they spurn that remote control when it is offered? Or is the accumulation of speed, along with the accumulation of variety, along with the accumulation of wealth, a one-way street in human cultural evolution?

Broadcasters have to worry about this, and they believe it means they must be more efficient than ever in their use of time. Just as the technology of remote control has made it possible for you to run from boredom without leaving the couch, the Nielsen technologies have made it possible for television programmers to detect the first glimmerings of ennui, apathy, and listlessness almost before you yourself become aware of them. A minute is an ocean. At NBC, John Miller, executive vice president of advertising and promotion and event programming, explains just how fine-grained the decision-making has become. "Every station looks at every second of air time and uses it to the best of their ability," he says. "We're all bound by the laws of physics. There are only 24 hours in a day and 60 minutes in an hour and 60 seconds in a minute. Everybody looks at their time with a microscope to get the best utilization they can. It is the only real estate we have." One piece of news turned up by NBC's research dismayed the programmers: as a typical show reached its end and the credits began to roll, one viewer in four, with a remote control presumably in hand, would give in to the urge to press the Channel Up or Channel Down button. A full 25 percent of the audience would start flipping around. That was clearly intolerable. A 25-percent drop in market share in return for gratifying the egos of the cast and crew? The NBC 2000 unit addressed this problem by creating what is known as the squeeze-and-tease: the credits are compressed into one-third of the screen (carefully tested for borderline readability) while the remaining two-thirds is used for "promotainment." You might see stars bantering about and around the peacock.

If you actually take in the screenwriter's name on the right *and* chuckle at the wisecrack on the left, you are multitasking in yet one more way. Anyway, every network has quickly adopted the same technique, because it is just enough, it seems, to hold your attention for the critical ten or thirty seconds that would otherwise loom before you like an eternity.

The network's time obsession has changed the basic structure of standard shows like the thirty-minute (twenty-three-minute, really) situation comedy. Network programmers feel they can no longer afford the batch of commercials that used to separate the end of one show from the beginning of the next. So those commercials have moved inside the shows, creating little islands of program at the beginning and the end, cut off by several minutes from the main body. Clever writers use these for stand-alone opening jokes and codas. "It's jokes and story right from the git go—jump in and go," says Skip Collector, editor of *Seinfeld*. "That kind of relates to our lifestyle and our pace, everybody's rushing and going and that's what we're going to do." *Seinfeld* was one show that used the split-screen closing credit time for a final joke, rather than give it up for promo-tainment. It also dispensed with the traditional half-minute or so of opening titles: Mary Tyler Moore throwing her hat in the air week after week, or Cosby's family dancing around. More and more sitcoms just start with running story and flash a three- to five-second art card with the name of the show.

At least the major networks still program their airtime around the quaint assumption that viewers will arrive on the hour and half-hour and stay more or less in place. Many cable-television channels have abandoned that idea. Like parents giving up on mealtime and leaving an assortment of snacks in the refrigerator, they design their programming for a perpetually restless clientele. E! Entertainment, for example, passes the minutes with a pastiche of clips, interviews, promotional tapes, and similar fare, all designed to be glittering enough to hold the attention of channel surfers whenever they happen to drop in. One of its features is "Talk Soup," a compilation of brief moments from other networks' talk shows, as if talk shows weren't already in sound-bite territory. We're reaching the level of distillation of an abridgement of a sampler of a *Reader's Digest.* Every meal a tasting menu. Sometimes the miniaturization is the joke. Nickelodeon's TV Land channel squeezed in "Sixty-Second Sitcoms," complete with opening and closing credits, a tiny commercial, and time for, on average, two gags.

All these channels fill the gaps that used to be dead air by playing instances of a new miniature art form: "promos," "opens," "bumpers," and "channel ID's." NBC alone commissions eight thousand different promos a year. They range from ten seconds to the "long form" two minutes, and they represent an astounding deployment of technical sophistication, products of a marriage between computers and the visual arts. In the early 1980s independent designers with new computer-graphics systems, a Paintbox and a Harry, could suddenly produce complex animated effects in an hour that had previously taken a full day. With the ability to compose effects frame by frame, to create multiple layers, images dissolving into new images, designers know that the viewer cannot always keep up. But they can't always help themselves. If the technology lets them add layers, they tend to add layers. Some of the power of these bits of video lies purely and simply in their speed—the length of time between cuts steadily decreasing, to the point that we routinely absorb sequences of shots lasting eight frames, a third of a second, or less. For someone creating a ten-second channel ID that will be seen over and over again, an effect that cannot be parsed on first sight by a typical couch-bound viewer is not necessarily a bad thing. Designers sometimes don't know or care whether the viewers will actually see a four-frame image. It's an impression. Maybe they'll see more on the next viewing. A flashed image can be like a subtle allusion in a long poem, resonating just below the threshold of comprehension.

MTV Zooms By

People who revile the evolution of a fast-paced and discontinuous cutting style—and, for that matter, people who like it—have a convenient three-letter shorthand for the principal villain: MTV. The most influential media consultant of modern times, Tony Schwartz, offers this doctrine of perception:

> The ear receives fleeting momentary vibrations, translates these bits of information into electronic nerve impulses, and sends them to the brain. The brain "hears" by registering the current vibration, recalling the previous vibrations, and expecting future ones. We never hear the continuum of sound we label as word, sentence, or paragraph. The continuum never exists at any single moment in time.

Schwartz put his theories to work in some of the most famous political spots of the last generation, from the watershed 1964 anti-Goldwater commercial—a girl counting daisy petals juxtaposed with a nuclear explosion—to the fast-cut "Read My Lips" commercial that damaged George Bush in 1992. Schwartz sits now amid a treasure-house of aging tapes and memorabilia on the first floor of his Manhattan town house. He was one of the inventors of the supercompressed video montage—a two- or three-minute bit of film combining hundreds of nearly subliminal images of, say, the year in review. When the Cable News Network was new, its founder, Ted Turner, wanted shorter commercials to match the brisk pace of his two-minute newscasts. The thirty-second commercial, a bold innovation that had swept dizzyingly across the networks in 1971, somehow no longer seemed quite so swift. Turner hired Schwartz, who took a set of thirty-second spots and cut them down to eight seconds, seven seconds, five seconds. Now Schwartz looks at his watch and says, "I could do a . . . let me see . . ."—apparently he is playing something back in his head—"three-second commercial that would outsell any of them." He feeds a cassette into one of a rack of videotape players and, sure enough, three-second commercials: one or two quick images plus catchphrase. "Got a headache? Come to Bufferin." "You can see why Cascade's the better buy. Try Cascade." "As long as you've been taking pictures, you've trusted them to one film."

War and Peace it wasn't. But now even Schwartz is complaining about his up-to-date colleagues: "They see the stuff that's on MTV and they imitate that."

At MTV, the creative decision-makers offer no apologies. A company fact sheet asserts, as a kind of slogan, "MTV zooms by in a blur while putting things in focus at the same time." Music Television began broadcasting in the summer of 1981, with the Buggles singing, appropriately enough, "Video Killed the Radio Star," followed by the Who, the Pretenders, Rod Stewart, and others in hybrid blends of music, images of musicians performing, and other rapidly intermixed images, real or surreal, related to the music or not, but always *cut to* the music. The basic MTV unit was a three-minute movie created around a song. You might have been forgiven for thinking it was meant as a sort of wallpaper, something to put on in the background when you didn't want to *watch* television. Wasn't it really a descendant of television's Yule Log, burning away eternally at Christmas before a fixed camera while carols played on the audio track? Certainly the music video was premised on short attention spans. It is a three-minute format within which no single shot is likely to last more than a second or two. MTV soon became one of the United States' foremost cultural exports, playing to 270 million households, including those reached by satellites over Southeast Asia, Mexico, and South America. Besides music videos—which evolved into a fantastically crisp and artful genre—the network has sent out its own talk shows, dance shows, pick-a-date game shows, and, most intriguingly, animated cartoons, like the famous, dim-witted, super-ironic Beavis and Butthead.

The not-so-hidden premise of Beavis and Butthead is that even music videos are slow-paced and boring, so you need an overlay of comic commentary. In their own way, though, Beavis and Butthead are painfully slow—MTV going conventional and letting story, rather than music, dictate the pace. The MTV animation style is deliberately static; it makes the typical Disney feature look like a madcap action film. The dialogue staggers along as if through mud, and the comedy relies heavily on reaction shots (so standardized that the animators call them by name: "Wide-Eyed 1," "Wide-Eyed 2," "This Sucks").

"We love pauses—pauses are like, hey!" says Yvette Kaplan, supervising director, as a bit of tape makes its way through the editing room, a segment involving an impotency clinic. "Oh, yeah," Butthead is saying in the sequence now running over and over again through the editor's screen. "Huh-huh. Me, too. Huh-huh. Maybe that place can help us score."

Of all the visual arts, animation takes the tightest control of every fraction of every second. On carefully diagrammed sheets, each consonant and vowel of each word is assigned to its precise one twenty-fourth of a second frame. The characters' mouth movements have been reduced to an essential grammar of just seven or eight basic positions, enough to cover all English speech. This particular joke strikes the team in the editing room as . . . slow. There seems to be a lag in the line. "The pacing is everything," Kaplan says. "When it's flowing, it's just safer—you don't have time to drift away and miss the humor." They delete the "me, too" and nudge the pace forward a bit more by overlapping the final fraction of a second of the sound track with the visual track for the next scene. Alternatively, they might have jumped to the next scene's dialogue before cutting away visually, or they might have started the music for the next scene early—clever pacing techniques that viewers have learned to interpret automatically and unconsciously.

"The audience has gotten more sophisticated and you can take certain leaps without people scratching their heads," says Abby Terkuhle, president of MTV animation. And of course, we're starting young. "It's intuitive," he says. "Our children are often not thinking about A, B, C. It's like, okay, I'm there, let's go! It's a certain nonlinear experience, perhaps."

Allegro ma Non Troppo

Once upon a time, before Music Television, before remote controls, before books on tape and Internet streaming media, a possible method of enjoying a basic art form was this: a person would sit down and listen to an entire symphony, for however long that took. It is not so easy anymore. Even people who love classical music find themselves bereft of the act of will necessary for blocking out forty uninterrupted minutes away from telephone or computer. Halfway through the adagio they feel a tickle somewhere between the temporal and occipital lobes and realize they are fighting an impulse to reach for a magazine—but no multitasking now, please! The great symphonies and concertos and operas and chamber pieces from 1700 onward, the works that make up the core repertory of classical music, were composed with the idea that listeners would be attending to them with all their conscious minds, having arranged their schedules and perhaps paid money to occupy a seat in a concert hall for a set time. Listeners were enclosed in ritual spaces, with nothing to do but listen and watch. Now there are other temptations.

Not everyone gives in, of course. Some people still lovingly find records at the local music library and sit for hours in the listening carrel. The classical-music aisles of record shops are more crowded than ever with new releases of standard and not-so-standard repertory. Still, there are signs. The liveliest category of classical-music recording—and for that matter every other kind of recording—is the compilation-anthology-sampler category: archives, remixes, collections, selections, celebrations, greatest hits, unreleased treasures, classics, best-of's, rest-of's, essentials, and favorites. There seems to be no limit to the number of orthogonal slices a record company can take through the same literature. *German Music for Trombones. Gay American Composers. Favorite Opera Choruses. Orchestral Excerpts for Tuba. Lovers' Greatest Hits. Melodies for Praying.* Scores of disks, in fact, titled just *Melodies.* Or *12 Hommages* or *13 Motets* or *14 Greatest Tenors* or *15 Nocturnes* or *16 Orthodox Stars* or *17 Jewels in the Crown of the Baroque* or *100 Fiedler Favorites* . . . the one certainty in all these compilations is that no single track will demand more than a few minutes' investment. There is no *Bruckner's Greatest Symphonies* compilation disk. But then again, record labels have managed to take snippets even of Bruckner, for *Music to Soothe the Soul, After Hours Classics, Classics for Relaxation, Horizons: A Musical Journey,* and more. Not to put too fine a point on it, you may simply wish to buy the disk called *Presto! World's Fastest Classics,* seventeen tiny opuses without context, from Goldberg Variation Number 26 to *Flight of the Bumblebee*, by way of an overture, a troika, an etude, one of *The Planets*, a sabre dance, and an allegro con brio.

The New York Philharmonic offers compressed "Rush Hour Concerts." Awareness of rush hour also permeates classical-music radio stations, such few as remain, and the art of programming has changed accordingly. Some stations let their "Music Director" software do the programming; others use it as a card catalogue, listing selections by duration. Given a choice between the James Judd *Eine kleine Nachtmusik,* at 21 minutes, 31 seconds, and the brisker Bruno Walter version at 20 minutes, 18 seconds, some programmers automatically choose to save the minute. For that matter, they are probably playing no more than the opening allegro. Stations that used to be run by purists now broadcast isolated movements from longer works—better a Mahler scherzo, they reason, than the same old seven-minute overture. Or they excise single movements, the adagio of a Vivaldi concerto or Haydn symphony, worried that music so slow to unfold its inner workings will lose the attention of a capricious drive-time audience. Sometimes stations trim the seconds-long breaks between movements—dead air—jerking the listener from the andante to the minuet without a breath.

With all the arts making their small sacrifices to hurriedness, music lovers can hardly expect to be immune. There is a special kind of pain, though. Music is the art form most clearly *about* time. The passing seconds are its canvas and its palette. There are extremes of slowness in music that stand out like peaks of the Himalayas—the *Heiliger Dankgesang* movement of Beethoven's Opus 132 string quartet, or the final heart-stopping adagio of Mahler's Ninth Symphony (Leonard Bernstein, conducting, took six minutes to crawl through the score's last page). These seemingly endless passages evoke death by information deprivation: when rich flurries of black-beamed chords give way to the simplest sustained single tones is precisely when Lewis Thomas, listening, found himself thinking of "death everywhere, the dying of everything, the end of humanity." Modern composers play with the extremes in different ways: Milton Babbitt creating a rapid structural complexity that even *Sesame Street* graduates must find hard to follow without practice; minimalists like Philip Glass letting a structural simplicity flirt with the edges of boredom. Either way, each composition means to carve a shape out of time. The composers implicitly ask you to isolate a certain amount of time as an experience. "We have a sense that time marches inexorably forward, and music defies all of that," says William Lutes, the program director responsible for classical music on Wisconsin Public Radio. "Music absolutely defies it. Music takes what was past and turns it into the future. Music expands the present moment . . ." But his average selection is down to seventeen or eighteen minutes—still longer than most stations', and longer, he fears, than many listeners will tolerate. "People are more inclined to grab something off the radio when they need it and come and go more quickly."

Can You See It?

Problem for the next generation: "Movies at a theater take FOREVER to watch—no fast forward," says a character in the Douglas Coupland novel *Microserfs*. "And VCR rental movies take forever to watch, even using the FFWD button."

Solution: "This incredible time-saving secret—foreign movies with subtitles! It's like the crack cocaine equivalent of movies." You can watch even an art film in less than an hour. "All you have to do is blast directly through to the subtitles, speed-read them, and then blip out the rest. It's so efficient it's scary." Then again, it shows how quickly we grow adept in the use of our tools for the manipulation of time. The villain of Martin Amis's 1989 novel *London Fields*, trying to enjoy choice bits of television as pornography, alternates fast-forward and slo-mo—every session a new work of art. When on-line entertainment writers say of a starlet's famous seduction scene, "Celebrity Profiles recommends renting this movie if you own a VCR with good freeze-frame," you are supposed to know what they mean. Your entertainment quiver is not full till it includes good fast forward and good freeze-frame.

Alternatively, when the pace of sights and sounds coming from the screen leaves us hungry, we cope by adding layers. Of course, we multitask, watch television and eat and leaf through a magazine and do needlepoint. But the programmers do not want to lose a percentage of our attention to needlepoint, so they fight back: they add the layers for us. VH1, a younger counterpart to MTV, created a hit with an art form called Pop Up Video: recycled music videos, in themselves now too slow or familiar to be captivating, overlaid with cartoon-style balloons as a side commentary on the main action. There can be dozens of them in a three-minute video, sometimes multiple balloons, containing quick jokes or historical facts or puns on the lyrics. Tina Turner sings "Missing You," and a balloon pops up near her bare neck to explain, "sternocleidomastoid muscle." Just in case you're bored. The Pop Up Video gimmick was quickly copied in commercials and other programming. In part, this is what we call irony. There is an incongruity of tone and attitude between the two layers of a Pop Up Video, the base layer and the balloon layer. The balloon layer mocks the original video, feeding off viewers' eagerness to smirk and wink and otherwise distance themselves from simple images that they accepted at face value just months before. Even apart from the irony, however, the new layer simply provides something *more*—a second perspective where a single perspective no longer suffices. It's not just recycling—as Woody Allen has it, "an adaptation of a sequel to a remake." It's cud-chewing, meta-entertainment—like the sideshows for people waiting in line at Disney World.

The finest example of multilayered meta-entertainment—recursive, self-conscious enrichment of the plodding movies of an ancient era—is *Mystery Science Theater 3000*, a television show created in 1988. The show runs decades-old B-movies in the background while three characters silhouetted at the bottom of the screen provide a running commentary of wisecracks. These bad old movies leave plenty of time to fill, by modern standards. The ironic voices are not shy about pointing out the sluggishness:

"I bet if these guys filmed *Citizen Kane* it would have had a twenty-minute sled sequence in it."

"Let's take a fifteen-minute break. But keep the camera rolling."

"Please remain seated until the movie grinds to a complete halt."

"Could something please just happen?"

Along with the jokes they slip in a sophisticated analysis of film technique, old and new. Audiences and filmmakers alike have learned how much can be omitted from simple narrative sequences. "When film was a new medium, they didn't know they could collapse time," says Michael J. Nelson, the show's head writer. "There was a lot of padding." We no longer need to see the man getting out of the car, and closing the car door, and walking up the steps, and knocking, and entering. The camera can jump from car to living room without leaving us behind.

So Nelson and his colleagues bring these plodding movies back to life with their comic transplants—yet as a viewer himself, he sometimes despairs. "I've been nearly driven to madness by the pace of commercials and television, so I watch much less," he says. "Not that I'm becoming a Luddite, but I'm withdrawing a little. It's style over substance."

Filmmakers have experimented with speed almost from the beginning. The first films were one long shot, in real time before there was anything *but* real time. By the 1920s, though, Sergei Eisenstein was pioneering techniques of fast cutting that seem radical even today. "Without even using a Movieola," says Walter Murch, who edited *The English Patient*—digitally, of course, on an Avid nonlinear system. "Eisenstein did it blind, like putting together pieces of cloth on a tailor's table." Today Eisenstein would have full-screen editing and playback at twenty-four or thirty frames per second, with thirty-two levels of undo/redo; a database for tracking footage; ready-made dissolves, wipes (diagonal, matrix, or sawtooth), flips and flops, blow-ups and resizes, peels and pushes, conceals and squeezes—and then there are color effects and motion effects (slow, fast, freeze-frame, reverse). Editors can drop single frames to create a subtly accelerated staccato feeling or just to bring a new segment in at the precise fifty-eight seconds required. All this technology has conspired to create breakneck production schedules in Hollywood and in newsrooms. Editors and directors differ about whether it has affected their art as well, but any comparison of older films (even through the 1970s) and newer ones shows an enormous difference in the length of typical shots and in the quickness of rapid-fire action sequences. In 1968 Stanley Kubrick's *2001: A Space Odyssey* seemed a reasonably adventurous movie; thirty years later, audiences conditioned by the space epics that followed could barely tolerate its endless, languid shots of moons and space stations drifting on screen to a classical-music background. Mitchell Stephens sees the same painful sluggishness in classic television documentaries he presents to his journalism students at New York University. "Fast cutting has arrived not as a tic or an affectation," he argues, "but because it makes video of all sorts more interesting to look at and potentially more informative." Films, commercials, and music videos that an earlier generation would have found incomprehensible or annoying can please you or challenge you with, as Stephens says, their "revelatory, narrative-challenging, character-debunking, seemingly illogical, unvirtuous, graceless, surreal quickness." If part of the effect is merely visceral—agitating, attention-grabbing, jerking the viewer out of slumber—that does not necessarily entail a loss of seriousness or meaning.

The English Patient was a famously languorous movie, but it needed an audience prepared for tricks of pace and time that might not have been possible even a few years ago. The film moves through forty different transitions forward and backward in time, into different people's memories. It uses a whole range of devices to pull the viewer along: visual dissolves and sound transitions, as when the rhythm of a key thrown on stone in a game of hopscotch weaves its way into the Arabic dance of the next scene. "The fact that we got away with that convoluted a temporal landscape is astonishing," Murch says. "Gradually we have found more and faster and better ways of articulating all that." We have come to understand speed. We may resent it as a substitute for suspense. (Hitchcock knew that suspense is *slow*; suspense is not the brilliantly fast-cut shower scene in *Psycho*; it is Cary Grant carrying a glowing glass of milk, possibly poisoned, up a flight of stairs that lasts, it seems, forever.)

We appreciate speed, as a tool of storytelling or just as a bright challenge to our senses. We admire speed, and always have, as raw virtuoso performance—Jascha Heifetz flashing through an encore piece, always teetering on the verge of breaking a string or flubbing a hemidemisemiquaver. True, allegros without adagios grow tiresome. Slow music can have its subtler kind of virtuosity, the weightlessness of a bicycle rider staying balanced while drifting to a halt. Gustav Mahler is supposed to have advised a young conductor: if you think you are boring your audience, slow down. If there is an ultimate limit to the pace of entertainment, we must now be approaching it, just as Olympic sprinters are approaching the human limit for the hundred-yard dash. In some ways, we are past the limit. Any day now, lawyers will take note of the considerable quantities of television text—including copyright notices and advertising fine print—that flash by too fast for any human to read it. The inexorable one-upmanship of movie action sequences has clearly left the laws of physics behind, and audiences are noticing.

The audiences, though, have themselves been altered. We are different creatures, psychologically speaking, from what we were a generation ago. "If you look at a one-minute commercial from the fifties," says Barry Levinson, "it seems forever. It seems so long it's like a show." Impatience that breaks out inside a minute-long time frame seems pathological. How much can we pack in, finally? Back over to VH1, just in time to catch this morning's ":60 Album Review." Yes, that's sixty seconds. In one minute a series of reviewers, with fast-changing graphics floating behind their heads, will discuss various new records. Also on screen, in case you don't already get it, is the blur of a digital timer ticking off the whole minute in *hundredths* of a second. It seems that we—we viewers of mass entertainment—have lost some of our ability to sit on the porch and daydream as the clouds float by. In the 1996 film *Lone Star,* the old-woman-on-porch defied her stereotype by whiling away the time with a hand-held electronic game machine. We know that these games are almost all about pure mental speed of one kind or another. As our attention has demanded more stimulation, we have gained an ability to process rapid and discontinuous visual images. It seems that we are quicker-witted—but have we, by way of compensation, traded away our capacity for deep concentration? No one knows for sure.

"We do suffer these days from a little bit of attention-deficit syndrome, whether it can be diagnosed or not," says Rick Wagonheim of R/Greenberg Associates, a leading creator of digital effects. "Are we smarter? Probably not. Are we able to absorb more information in a short time? Probably." Like it or not, commercials, combining twenty or thirty or more individual shots in as many seconds, are a cauldron of new techniques. Perhaps because commercials are video with a clear purpose, perhaps because they are video with a nearly limitless budget, they display the most gripping and exciting styles, from miniature storytelling to quick-change manipulation of our emotions. It's no wonder that so many film directors and editors are emerging with a background in both commercials and music videos: from Michael Bay, for example, director of *The Rock,* and Hank Corwin, editor of *Natural Born Killers* and other Oliver Stone films. Corwin, whose experimental style has pained some critics, dismisses them:

> Fuck 'em. They're stupid. They'd better get with it, because that's the way of the world. I've been tagged unfortunately as like the MTV-style guy, and that's sort of pejorative. There's a lot of crap out there, and you can't disregard that. But we're going into the millennium, and things are always moving, things are always changing, things are very kinetic.

It's almost brain chemistry. We're more sophisticated. We're like fighter pilots doing a panel scan, absorbing data from all our instruments at once . . . multitasking . . . in real time. "Let's give ourselves credit," suggests Stephens.

> We have learned to grasp quickly. We can read signs, change lanes and avoid other vehicles at seventy miles per hour while also listening to a song and planning our weekend. . . . Things come at us at a rate our ancestors could not have imagined, and we handle them.

"Our eye has quickened," says Michael Elliot of Mad River Post, who produced groundbreaking commercials for Compaq, MCI, Reebok, Epson, and others—commercials with fast pace as a spoken theme as well as a technique. As you watch the quick montages, the shotgun blasts of views from scattered angles, you can't help but notice the pathos of the soundtracks, catering to what advertisers see as the deep concern of the audience. You hear the voices of men and women, at home and in the workplace, talking about their hectic lives, their need for time-saving, their hunger for speed, their fear of overload. It hardly matters what product: fast computer or fast telephone service or fast athletic shoes. The words and images flash by, because you, the viewer, have adapted to the blur. A more stately and deliberate tempo would frustrate you now. With no irony the message comes through: you have too little time, and you are working too hard, so buy this—quick.

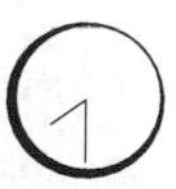

High-Pressure Minutes

Oh, the buzzing of the bees in the cigarette trees. The great hobo ballad "Big Rock Candy Mountain" contained a few key passages that are generally left out of recordings for children: for example, "I'm bound to stay where you sleep all day / Where they hung the jerk that invented work." We have mixed feelings about work and leisure, to put it mildly. Article 23 of the Universal Declaration of Human Rights says that everyone has the right to work. Then, Article 24 says that everyone has the right to rest and leisure. "Whoever has to work for a living," declares Sebastian de Grazia, "is blocked on the road to wisdom and suffers, as far as leisure is concerned, the fate of slaves." Our best seers had hoped for a different life. Utopia was not supposed to be a place where people worked. Anyway, work was supposed to be something *other people* did. "The idea that the poor should have leisure has always been shocking to the rich," Bertrand Russell wrote acidly in 1932. "In England, in the early nineteenth century, fifteen hours was the ordinary day's work for a man; children sometimes did as much, and very commonly did twelve hours a day. When meddlesome busybodies suggested that perhaps these hours were rather long, they were told that work kept adults from drink and children from mischief." He looked forward to a future where a full day's work would be four hours.

Instead, movie-makers, television producers, journalists, building contractors—in so many industries people are working faster. Their work may not flow on visible assembly lines, but their invisible assembly lines are accelerating. The gears and the passing belt are driven by faxes and E-mail messages—time-saving forms of communication. Without the snail's pace of ordinary mail serving as friction, their projects slide forward in a rush. Also driving the work is the expectation of its quickening, plugged-in consumers. You do demand timeliness in your news and your movies and your legal work, don't you? Some jobs fill the seconds and minutes of every hour with the increasing density of transistors on a silicon wafer. Air-traffic controllers, boiler-room salespeople, workers in the computer-coordinated factory—any worker, for that matter, whose output is monitored by computer—all these job categories work hours more rigid and unyielding than the hours in a university department or a law firm. There are pounds of lead and there are pounds of feathers. Jobs involving telephones tend to fill hours with a special ruthlessness.

"It was like standing at the beach and trying to stop the water from coming in," says John Bonano, a telephone company executive, of his brief experience as an operator providing directory assistance. He is standing with a group of directory-assistance experts—all waiting exasperatedly, in fact, for a slow elevator deep inside a secure telephone-company building in midtown New York. They are heading for the central room where their team of operators—seventy-two men and women at peak periods—sit in gray cubicles and answer the 411 calls for all Manhattan and parts of Brooklyn and the Bronx. It is a process that takes time-saving to its modern extreme. In the metropolitan New York area the telephone company handles 5.5 million directory-assistance calls a day; with that huge number as a multiplicand, any saving in the few seconds devoted to each call releases years of human life back into the cosmic total of free time.

This place is a laboratory for the intensification of the modern work experience and for the most extreme reaches of time-saving technology. The operators, having learned to decode more accents than any group of United Nations interpreters, having learned a standard mnemonic alphabet (M as in Mary, P as in Peter . . .), having donned headsets and readied their hands above specialized keyboards, set out each day to beat the department average: twenty-one or twenty-two seconds per call. The best claim a personal average of sixteen seconds—meaning that during a stretch of one hour and forty-five minutes they handle 394 calls, before taking a fifteen-minute break or a half-hour lunch and putting the headset back on.

Twenty-one seconds for the operator does not mean twenty-one seconds for the caller. This is no longer a real-time experience, where one person calls one operator and they have a conversation. Technology has severed the link between caller and operator—or skewed it, anyway, digitizing and time-shifting their voices. Your call to directory assistance and directory assistance's call from you pass in parallel universes. First, beginning in the 1980s, the operators stopped speaking the digits of the phone numbers they found; computers do that. Specifically, the operator presses a key labeled Audio Release, and far away, in the depopulated, air-conditioned clean room of the central-office switch, an electronic circuit in a many-port cabinet the size of a refrigerator produces pulses that your telephone handset converts to the sound of spoken digits. That saves almost five seconds. But it doesn't save *you* five seconds.

While you listen to the number, the operator is already well into someone else's request. Similarly, at the beginning of the call, another cabinet reproduces the voice of, say, James Earl Jones, eerily pulsing the phrase "Welcome to Bell Atlantic" millions of times daily, followed instantly by yet another cabinet, far away, with a phrase like: "Directory assistance. What listing?"

It used to be "What listing *please*," but every millisecond counts. Impatient callers tended to break into the *please* with their request anyway, like baseball fans beginning their post-anthem applause on "land of the free" instead of waiting, as in past times, for "home of the brave." In some places the electronic voice asks, "What city?" but not in New York; New Yorkers know what city.

Your turn to speak. You say, "Motor Vehicles" or "Domino's, um, you know, Pizza" (to directory assistance, it is clear that the world lives on pizza and Chinese food; some keyboards actually have a dedicated *pizza* button). As you finish, your words are routed into the headset of the next available operator—who, while you were speaking, was still busy helping a previous customer.

What the operator hears and what you said just a moment before are not exactly the same. To save even more time, computers send your words first through software that removes any pauses or silences that might have marred your otherwise crisp delivery. This software tries to remove the uh's. It can also speed up the playback ever so slightly. The telephone companies would like to find the ideal setting for this compression: fast enough to save time, yet not so fast as to stupefy the operators. A workstation in yet another clean room monitors these variables: Compression Savings and Silence Removal Savings. It also monitors your performance—yes, when you dial in for directory assistance, you can perform well or poorly. The machine keeps a count of your Spoke Too Soon Errors (meaning that you lacked the patience to wait for the "what listing?" prompt), Spoke Too Long Errors ("the, um, good morning, can I have the, uh, number, for, like . . ."), No Speech Detected (you completely froze, or you went away), and other measures of imperfection.

If you do your job perfectly—manage to state, not too soon and not too long, in Standard English, an unambiguous request that produces a single listing on the console—then you will not be connected to the operator at all. The human role in directory assistance will be fully removed to the background. Intricate though the process has become, its engineers know that they have only reached a way station. Someday, they hope, voice recognition will save even more time. They are experimenting with automated recognition of a few very common phrases, like "Domino's Pizza," but computers still have trouble distinguishing yes and no, in the rough-and-ready polyglot environment of directory assistance. Meanwhile, when the telephone-company time-saving arithmetic reaches the bottom line, it appears that the new technologies save the company a few seconds, while *costing you* a few seconds, on average. They actually shift a bit of time, that is, from your ledger to the company's. On average.

That is partly your own fault. You could save time, too, if you learned that pressing the "pound" key lets you skip past the automated prompts. But you haven't learned that, have you? You didn't take the time to read the instructions on the insert enclosed with your telephone bill.

Not long ago, the operator sat flipping the pages of giant telephone books. Then came microfilm, briefly. Then computer consoles. The human brain, which so far cannot be replaced, can at least be pushed to its top speed. Electronic tones in the headset signal each incoming request. "Staples on 57th Street, please?" Find the listing, move the cursor, hit Audio Release to let the computer take over. Tone. "Brooklyn Union Gas?" Cursor, Audio Release, tone. "Maimonides Hospital?" A trained operator types MA MED, moves the cursor, and hits Audio Release before the untrained brain has even begun to parse *my-mon-a-deez.* It seems brutal—Sisyphus in fast motion, with a new stone to roll every twenty-one seconds—and for some, it is too much. But some operators say they thrive on the rush. They try to improve their personal averages; their minds find just the right shortcuts; their fingers fly independent of conscious control. And they multitask: they are able to detach a separate piece of their mind for daydreaming, even as the tone comes again.

It's the same with the ultimate in fast-paced high-pressure jobs, the one where humans at their consoles control objects moving five miles per minute and where "crash and burn" is not a figure of speech. The most stressful single control room in the industrial world may be the New York Tracon in Westbury, Long Island, responsible for seven thousand flights a day. "The controllers curse and twitch like a gathering of Tourette sufferers," the writer Darcy Frey observed in 1996, "as they try to keep themselves from going down the pipes." The burnout, the stomach acid, the breakdowns are legendary. But the reality that keeps the system going is that even more controllers genuinely like the pressure. They find it a form of mental athleticism, with grace and virtuosity, knowledge and power. They command a special language of instantaneity. Just as languages of the polar region have words for the many varieties of snow, so the air-traffic controllers master the various nuances, specified in the Federal Aviation Regulations, of *immediately, no delay, expedite,* and *urgency* (a condition "requiring timely but not immediate assistance"). There are also less orthodox ways of specifying urgency. "Hey, you're in New York, buddy," a controller named Tom Zaccheo informs one aircraft. "I need you to descend in a New York minute, not a hillbilly minute." Their type is Type A. When they drink, it's coffee. When they eat, it's takeout Chinese. One of Frey's inadvertent twitchers was the 1995 air-traffic control overtime champion, Jim Hunter. "I'm sure there's long-term effects of working so much traffic," Hunter said, his leg jiggling. "Actually, I get a buzz off it. It's true." Like a drug without the actual chemicals, and the busier the better.

Time and Motion

You're not directing air traffic. In point of fact, you're just baking potatoes. Still, there's no reason you can't do it briskly.

Mary and Russel Wright, in their 1950 *Guide to Easier Living,* pointed out some common time-wasting errors modern housewives make in the course of baking potatoes. They store the potatoes too far from the sink. They turn on the water and then reach for the brush, instead of using both hands to perform these steps simultaneously. They carry the potatoes to the oven one or two at a time, when they could put them all in a pan and make one trip. Mistakes like these—multiplied a thousand times and combined with a fussy, snobbish style of decorating and entertaining, handed down from the English manor house—left the average American housewife with a staggering sixty- to eighty-hour work week. "Eliminate unnecessary steps and motions," the Wrights exhorted. "Combine, rearrange, make them easier. The desired result is housekeeping minus all that is unnecessary, unduly arduous, and time-consuming." In other words, perform a rigorous time-and-motion study of the kind an efficiency expert would bring to an industrial corporation. No task is too small for the application of "science." The housewife must think of herself as the home's production engineer. Case in point:

STEP	MOTION (details)	TIME
Goes to range	Turns on heat	Time begun 3:44
Goes to cupboard	Opens door (right hand)	
	Takes out pan (left)	
	Transfers to right hand	
(Goes to sink, oven, etc.)	(etc.)	Time ended 3:53
TOTAL STEPS:	TOTAL MOTIONS:	TOTAL TIME: 9 MIN.

Not exactly Zen and the Art of Baking Potatoes. Aware as you are of the peril of hurry sickness, you may recoil at the idea of bringing a stopwatch to bear on every trivial task. Anyway, what do you care? You have the microwave, you have prewashed potatoes, you have Tater Tots.

But the Wrights understood the danger. Their goal was time-saving without hurriedness. If we have come to wonder whether we can have one without the other, it is because we have already internalized the time-and-motion philosophy. Not only do we multitask, but we apply sophisticated critical-path scheduling algorithms to the second-by-second minutiae of daily life. Consciously or unconsciously, we plan the next five minutes in the kitchen with the kind of rigor formerly reserved for yearlong construction projects. Consider Robert Otani of Los Angeles:

> I map out what I need to get done: Start the PPP sequence, but launch the E-mail program first so it loads up while the modem beeps, whines and shushes, then download the software updates, then check the newsgroups.

Once that's started, I can make my coffee: insert grinds into machine, turn machine on, while the coffee is brewing, start the toast and cook the eggs.

If I don't brew before I cook the eggs, then I've wasted time!

Otani making breakfast is the master of his fate; so were the housewives to whom the Wrights addressed their manifesto for an *easier* life. The workers whose toil was the first object of time-and-motion studies did not always welcome the principles of scientific management so willingly.

The efficiency expert's stopwatch—and it was a special stopwatch, its face divided into tenths and hundredths of a minute for ease of calculating—arrived in the workplace in the last years of the nineteenth century in the hands of Frederick W. Taylor. (Sometimes he hid the stopwatch in a hollowed-out book.) Perhaps this was an inevitable transition in the evolution of industrial production: with the world's economy resting more and more on competition between manufacturing enterprises, someone had to notice that the key variable in the arithmetic of production was always time. Tons of pig iron loaded on a freight car per day. Cords of wood milled per hour. Feet of steel cut on a lathe per minute. The calculus of productivity, *anything* per unit time, is so deeply engrained in the post-industrial world that we can barely conceive of a workplace psychology omitting it. Yet it did not exist before "Speedy Taylor" forged his methods and ideas in the factories of the Northeast in the 1870s, as the Industrial Revolution reached its height. Taylorism is the ideal of efficiency applied to production as a scientific method—humans and machines working together, at maximum speed, with clockwork rationality.

One product of Taylor's obsession with maximizing efficiency was the invention of "high-speed steel"—a leap forward in the process of cutting forged metal on a lathe, patented in the closing weeks of 1899. The contribution of high-speed steel to the accelerating prosperity of the modern world is hard to comprehend fully a century later. The Bethlehem Steel Company demonstrated it at the Paris Exposition. For those who saw the red-hot tool slicing cylinders of steel, it was the moment, as Taylor's biographer Robert Kanigel puts it, "when they watched the world speed up before their eyes." In his quest for ever more efficiency, Taylor barely distinguished between the lathes and the men who worked them. His treatise on, simply, "Shop Management" four years later, in 1903, made a powerful mark on industrial leaders then and later. It was a farewell to an unhurried world. "After the men acquiesce in the new order of things," he wrote, "it will take time for them to change from their old easy-going ways to a higher rate of speed, and to learn to stay steadily at their work, think ahead, and make every minute count."

Yes, we have learned to make every minute count, and a good thing, too, because Taylor warned that those who were too stupid or stubborn to speed up would have to drop out.

> In reaching the final high rate of speed which shall be steadily maintained, the broad fact should be realized that the men must pass through several distinct phases, rising from one pace of efficiency to another.

Legions of efficiency experts, management consultants, and industrial psychologists have followed, displaying "Work Smarter Not Harder" placards on their desks, offering Gantt charts and learning curves and standard operating procedures. Taylor's disciples Frank and Lillian Gilbreth adapted the early motion-picture technology to create *chronocyclegraphs*—as Frank said, "to record the time and path of individual motions to the thousandth of a minute." Lillian proposed a new kind of manager: the Speed Boss. Today's speed bosses use software to assemble time-and-motion studies from standardized parts: get tool, place tool, focus eyes. We don't have to like it. We can assign Taylor ultimate blame for the creation of what the psychologist Robert Levine calls "Ticktockman" (you know who you are). Levine cites a slightly implausible time chart from the Systems and Procedures Association of America, one of many such institutions that owed their birth to Taylor:

> Open and close file drawer, no selection = .04 seconds;
> Desk, open center drawer = .026 seconds;
> Close center drawer = .027 seconds;
> Get up from chair = .033 seconds;
> Sit down in chair = .033 seconds;
> Turn in swivel chair = .009 seconds . . .

If you own a stopwatch, you want to use it. And when you know that you are swiveling in your chair in a mere nine milliseconds, who could blame you for wondering whether you could improve that record?

In the world's Taylorized factories, assembly-line efficiency is by its nature brutal, stripping craftsmen of autonomy, overriding what might have been a more natural, variable work rhythm. Then again, the countless little speed-ups engrained by efficiency experts in every facet of the modern workplace have created wealth and brought prosperity, or so economists will argue. "Each day we reap the material benefits of the cult of workplace efficiency that he championed," writes Kanigel, "yet we chafe—we scream, we howl, we protest—at the psychic chains in which it grips us."

The Paradox of Efficiency

At daybreak on a Wednesday in March, a McDonnell Douglas Super 80, No. 241 in the American Airlines fleet, takes off from Phoenix for Dallas–Fort Worth. A quick stop, and the plane continues to Richmond, Virginia, and then to Norfolk, and then back to Dallas. A generation ago, when airline scheduling was performed on big sheets of paper by men wearing green eye shades, that would already have been an unusual combination of cities for a single plane in a single day. An airplane flying from Phoenix to Dallas would most likely have returned directly to Phoenix. But the trek of No. 241, now back at the Dallas hub for the second time that day, turns more bizarre with a northward excursion to Calgary, Canada.

The next day, the jet returns to Dallas before flying to Los Angeles and then back east to Austin.

The next day: Austin to San Jose to Dallas to Nashville to Chicago to Denver.

Denver to Chicago to Boston to Chicago to Tampa.

Tampa to Chicago to Dallas to Chicago to Dallas to Des Moines.

Des Moines to Dallas, and now, nearby, as the staff of American's cavernous System Operations Control Center converse softly before large-screen workstations and eat takeout lunches at their desks, the computers show No. 241 in the air, en route to yet another city, San Diego, its fifteenth destination that week.

Its ramblings are not random; they are precisely charted by computers. The goal is a schedule of maximal efficiency—the best, or near-best, of the quadrillions of possible solutions. Scott Nason, the airline's chief information officer, tracing the past and future peregrination of this one aircraft on his console, guesses that the pattern of destinations and layovers has grown so tangled and involved that it will never repeat itself. All this complexity has a purpose: the saving of minutes. Presumably the minutes add up. Nason says, "Some of the minutes are very important."

He walks across the darkened command center, where all of American's division chiefs will gather at computer stations in the event of crisis—strike, war, hurricane. Through the window that makes up one side of the room, he looks out over the much larger control room below. "We can lock the doors and from here we can run the airline," he says. The big room is most fundamentally a computer room, too—those human beings, with takeout lunches next to their keyboards, are mostly there to monitor a vast calculation machine tying together data streams not just from every ticket counter and every airport gate but direct from electronic sensors in the doors, wheels, and brakes of every jetliner. The networking of the modern world finds expression here: the free-flowing connections between devices, calculating machines, display screens, and human overseers control virtually everything that needs to be controlled.

The sensors have four basic messages to send—Out (from the gate), Off (the runway), On, In—and right now Frank Botti, who is running the center from a many-screen console, is thinking about the "pineapple DC-10" that should have been Out and Off for Honolulu three hours earlier. He can assume that would-be vacationers are boiling with frustration on the ground in Chicago, where the jet is sitting with an engine failure. Botti barely glances at his maps, his flight list, more maps, the giant floor-standing display of National Lightning Detection. In a less efficient era, the waste of simple back-and-forth scheduling might have meant an extra aircraft or two just waiting idly, costing the airline money, but luckily available to fill in for the out-of-service DC-10. Now, with scheduling approaching perfection, less than 2 percent of American's fleet lies fallow at any given moment. So the nearest replacement plane happens to be in Dallas. A crew must fly it to Chicago. And another crew-scheduling problem is developing as the minutes tick by. The pilots have been sitting for quite a while. This time counts as time on duty, and now the long flight to Hawaii would push them over the legal maximum. So another crew must be found to replace them. Those vacationers will wait longer, and they will never know exactly why.

This is the paradox of efficiency. Air travel, like other intricate modern institutions, is a web of time and motion. Running parallel to the scheduling of aircraft runs a separate set of schedules for pilots and flight attendants, even more complex, determined by a mix of human and regulatory requirements. Alongside that, the computers continually recalculate aircraft weight and balance. Alongside that, they attempt to find increasingly efficient routing, using a real-time winds-aloft database in four dimensions: latitude, longitude, altitude, and time. They attempt to blend these facts of weather, chaotic yet pure, with muddier necessities: avoiding restricted military airspace; ensuring that a safe landing spot is always within reach, even on a single engine; negotiating with

Federal Aviation Administration controllers over their preferred routes.

With no other complications, the variation in winds aloft would be enough to destroy the possibility of precision in the most important statistic for each flight in the schedule: its "block time," gate to gate, flying time plus ground time. Block times are published and reported to regulatory authorities to the nearest minute. They are largely a fiction. On any given day the prevailing winds cause far more deviation in flight times than could any refinement in engine design or airfoil surfaces. A fast jetstream during a transcontinental flight can add or subtract an hour. Even at the perfect airline—an airline where arrival gates were always ready, where baggage-handlers never faltered, where flights were never overbooked—precise block times would be an impossibility. The reality of flying times is a fuzzy collection of probabilities: a statistical spread. Perhaps a flight would take 90 minutes a quarter of the time, 100 minutes another quarter of the time, 110 minutes another quarter of the time . . . what block time should a scheduler announce to the public? Corporate considerations further cloud the issue. In the seventies and eighties, airlines deliberately distorted these numbers by publishing unrealistically short block times. Then in the nineties, industry deregulation led to far less competition on most routes, so airlines began publishing unrealistically *long* block times—to improve their on-time performance records. They could even adjust the numbers so as to trade on-time performance in a less critical market for on-time performance in a market where competition was especially fierce. If an airline's on-time performance lagged in one fiscal quarter, easing the block times could help in the next.

These distortions aside, the airlines have truly succeeded in getting faster. As elsewhere in the delicate texture of modern life, time-saving has come more from the tautening net of efficiency than from raw speed. Airplanes themselves are not really speeding up anymore. The first commercial supersonic airliner, the Concorde, with its drooping needle nose and elegant delta wing, first carried passengers in 1976, cutting the New York to Paris time by half. British Airways had estimated that four hundred Concordes would be sold by 1980, and what a boon to business that would be! "We are now near the time when again we cut our traveling time in half," predicted *Nation's Business* in 1969. "The vehicle will be the supersonic transport—the SST—and the main beneficiary will be the American businessman. Legions oppose the SST on grounds it is too costly, too noisy, too complicated, too limited in usage. The Wright brothers heard those charges too, but then went ahead . . ." Seers imagined the next step forward—rocket planes arching through demiorbit from Tokyo to New York.

But these anticipatory paeans to the supersonic transport marked an ending, not a beginning. A quarter century later, just thirteen aging Concordes remained in service with Air France and British Airways. An American supersonic transport project was long since canceled and forgotten, and the Russian Tupolev was grounded. Partly to blame were the environmental curses of these planes—scarring of the ozone layer and unlikable noise twelve miles below their flight path. Routes were limited to the transatlantic run. The Concorde could not fly cross-country because of noise, and it lacked the fuel capacity to cross the Pacific. Perhaps most damning, though, was a variation on the Law of Diminishing Returns. There are diminishing returns in time-saving. The minutes saved blasting through the rarefied air of sixty thousand feet were so easily lost at the tollbooths of the Queens-Midtown Tunnel, lost again in the perpetual traffic jam on the Van Wyck Expressway, and lost again in the waiting line for customs clearance. (And every source of delay could be another business opportunity: IBM's vision of the future is its "Fastgate" system, which promises to cut your wait at immigration checkpoints to fifteen seconds. All you have to do in return for this savings is submit your "biometrics"—fingerprints and voiceprints—and other personal data for use in a state-of-the-art security database.) Airline marketing planners had actually imagined that business executives would fly from New York to London for lunch and back again that afternoon. Four airport taxi rides in a single day? Meanwhile, overseas telephone calls and E-mail and virtual conferencing became cheap and easy—real-time communication enough. Supersonic travel never found enough time-hungry travelers to become profitable. Space agencies and aircraft manufacturers kept higher-tech plans on their drawing boards, tempted by new materials and technologies, but the innovative commercial aircraft of the nineties were mostly commuter planes, slow, often with propellers, extending the web of service to small, out-of-the-way destinations.

So clean the planes faster; board by row numbers; deploy electronic ticket readers to help sort out the last-second seat-assignment conflicts that plague overbooked flights. Push the FAA toward a new system of Free Flight, in which aircraft could vary their paths to take advantage of the vicissitudes of winds and traffic, instead of following archaic routes based on the ground locations of old-style radio beacons. These are the time-saving efficiencies that matter now. In their own ways, time-hungry fliers try to maximize their own efficiency by becoming masters of the details. The most experienced travelers trade information about which airline terminals come first when you enter Kennedy International Airport by highway and which is nearest the heliport; which times of day and which days of the week have the most congestion; which gates are nearest the baggage claim area. They learn the euphemistic meanings of terms like "direct" when applied to flight schedules—direct flights are those with extra stops. They learn to sit near the front of the plane so that they can be among the first off; above all they want to avoid the horror of the oblivious slow-moving passenger who blocks the lone aisle, trapping a planeload of time-aware travelers. Of course they never check baggage if they can help it. Their obsession with time may or may not mean that they race to the gate at the last minute.

Some cannot stand the risk and pressure; others cannot stand the waste of sitting unnecessarily in an airport lounge, even with laptop and cell phone. Scott Nason has a strong preference for arriving early. He still remembers when he called down to make sure the 1:06 flight to Boston was on time and was told: "There is no 1:06. Flight 106 leaves at 12:54." He just got to the gate at 12:54—sweating.

The paradox of efficiency means that as the web tightens it grows more vulnerable to small disturbances—disruptions and delays that can cascade through the system for days. For example: American Flight 1128, inbound from Mexico, is now forty-four minutes late, and the computers are deciding whether to delay some of the connecting flights those passengers will be racing toward. This, too, will be a real-time decision based on complex modeling. The computer will know how many people are how many minutes late for each flight. It will consider the distance to the gate, the time before the next available flight to the same city, the likelihood of new delays at the other end. It will consider the passengers, too—if they have paid for first-class tickets, they will be more likely to find the gate waiting open for them. Pilots often accuse Nason and his computers of being overly fixated on time. "They ask, how can you close the door on a passenger running from three gates down?" says Nason. "Well, there are 130 people on this airplane looking at their watches."

It happens that Flight 1128 left Mexico late for reasons of "crew legality." The night before, its flight attendants, the only ones available, were twenty-seven minutes late leaving Miami and then forty-one minutes late arriving in Mexico. That delay cut into their legally mandated overnight rest period. So this morning the Dallas-bound flight could not depart until the precise minute when their rest period expired.

Networks like this are said to be tightly coupled. A complex construction project with a timeline scheduled with perfect efficiency, all the slack squeezed out of it, may be tightly coupled and a candidate for serious disruption. In the most extreme case, everything depends on everything else. Vibrations anywhere can be felt everywhere. The shin bone connected to the knee bone: that is tight coupling in the engineer's sense, especially if the ligaments do not allow too much flex. Charles Perrow, in his study *Normal Accidents,* extended the concept to complex systems where the coupling connects not physical parts but abstract services, people, and organizations. "Loosely coupled systems, whether for good or ill, can incorporate shocks and failures and pressures for change without destabilization," he notes. "Tightly coupled systems will respond more quickly to these perturbations, but the response may be disastrous." In tightly coupled systems, the connective tissue is often time itself. Process B in a drug-company production line or an aircraft-assembly plant or even a trade-school education must follow Process A as tightly as a ratchet and pawl. Waiting time or stand-by time can mean flexibility or safety. A tight system squeezes it out.

"Effects do cascade," Nason acknowledges. "We try to build in enough slack to break the cascades. We try to watch for cascades and truncate them. But some things you can't avoid." The hub-and-spoke system, itself a paragon of efficiency, with flights tightly scheduled in and out of a central focal point like Dallas, creates a particularly centralized site of vulnerability: a storm at the hub will cause delays nationwide. The system evolved because of welcome interactions between flights. The flights into and out of the hub support one another. Before hubs and before computers, there could never have been a regular flight from Shreveport to Portland or from Shreveport to Tokyo, yet now marginal cities like Shreveport join the network because these interactions add up to make the connections economically feasible: the Shreveport-Dallas flight strengthens the Dallas-Portland and Dallas-Tokyo flights. These same interactions, though, can send calamity racing across the system.

It all seems out of control—or rather, in control and yet out of reach, for us humans. In countless small ways, we seek to smooth the inefficient edges in our own lives. We have learned to keep efficiency in mind as a goal, which means that we drive ourselves hard. "Lost ground can be regained—lost time never," said Franklin D. Roosevelt in 1942, exhorting the nation to faster and more efficient weapons production, and he added, "Slowness has never been an American characteristic." A magazine advertisement running that year boasted that air transport "not only saves, but also gains days and weeks of precious time" and helps relieve "the greatest shortage of all—TIME itself." Taylorism, of course, had triumphed. Still, at mid-century a typical business would keep on a few over-the-hill workers in harmless nonjobs, would overlook an occasional late-afternoon card game in the office, would tolerate the routine two- or three-martini lunch. Not any more. All these inefficiencies represented slack that could be pulled in for a crisis, just like an extra DC-3 idling at O'Hare. We eliminate slack now. Ready access to information makes it easy. Six terabytes of data swim about in American's computer system at any given moment. When flights are canceled, the decision-maker is a workstation over in one corner, now called the Cancellator, formerly known as the Hub Slasher. There is no going back. The problems are too complicated. Everything would have to slow down.

365 Ways to Save Time

The pressure to maximize efficiency and save time in our everyday lives can become just as great, in its way. There are just not enough hours in the day. So here's some advice.

If you drive a lot, make use of the time by listening to audiotapes. It doesn't matter whether you're driving for work or for vacation. "You can learn a foreign language, improve your selling skills, quit smoking, and much more." This is tip No. 143 from a 1992 book, *365 Ways to Save Time,* published, appropriately, under the Time Warner Quick Reads imprint. Its author followed up soon afterward with *365 Ways to Save Time with Kids.* There is evidently no shortage of ways to save time.

But you, a careful reader, may be wondering whether multitasking in cars is really such a good idea. You are already flipping back through this same book to time-saving tip No. 52:

> Do you remember hearing the adage "To save time, do two or three things at once?" Times change. The wisdom gleaned from so much frenetic activity—and the resulting burnout and slipshod quality—is now: "Do one thing at a time and do it well."

You're listening, and yes, if you can reconcile the contradiction between this pair of tips, you will strive to achieve a single-minded concentration on quality. You will nevertheless want to practice tip No. 60: "You can sometimes sit quietly and do nothing. Go where your daydreams take you." Bearing in mind the apparent contradiction with tip No. 1: "Your time is a valuable resource. Don't waste it." To sum up: pack your driving time with extra mental stimulus, but don't take on more than one activity at a time, and for that matter don't forget to sit and do nothing. Empty time can be good—but come to think of it, you shouldn't have empty time, because you are practicing tip No. 70: "When you identify some uncommitted time, choose a task from your 'to do' list and make an appointment to do it." Unless, that is, you remember tip No. 25: make your to-do lists shorter. And for the sake of completeness, No. 27: make a "not to do" list.

Lists and files are the buckshot in the guns of time-management gurus. You are advised to keep lists of items to bring while traveling; files of receipts for tax time; files of credit-card slips to check against your bills; lists of scheduled recreational breaks; files of directions to destinations; lists of "everything you *don't* eat" and "what you *don't* spend"; lists of treats you might want to give yourself; lists of your friends' clothing sizes; lists of presents you have given; files of warranty cards; lists of books to read; lists of credit-card numbers; files of recipes; lists of dinner-party tasks; files of keys; lists of projects and then lists of the subtasks within each project; files of documents recording your personal history; lists of your debts; lists of questions to ask specialists; master lists, agendas, expense charts, and daily planners. In moments of clarity, these experts note that the entire filing enterprise is one of the world's greatest time sinks. A factoid much bruited in the time-management business holds that 95 percent of all documents ever filed remain filed forever, in the eternal file-folder night, never to be looked at again. But filing will not go away. One time-saving book devotes thousands of words to the care and feeding of file folders, topic by topic: Use File Folders, Collate Your File Folders, Label Your File Folders, Write File Labels by Hand, Use New File Folders, Use Expandable File Pockets, Organize Your File Folders, Organize Your Master File Drawer, Remove Your Hanging Files, and Empty Your File Drawer. Another proposes a system of red, blue, yellow, orange, and green folders, each with "running summaries" of their contents stapled to their inside covers—all this in addition to a planner/organizer and a separate box marked "file." Another devotes an entire chapter to "Ten Tips for Organizing Your Hard Drive." Thou shalt index thy files, for the sake of organization and quick retrieval. But don't forget to empty your in-box and purge six files, any six files, every day.

How much time can a person devote to time-saving?

When people come to you with demands on your time, don't say maybe. "Try to say yes or no whenever you can" (tip No. 306). Then again, don't say yes or no right away: "Say, 'I need twenty-four hours to think it over' " (tip No. 20). Then again, don't say yes at all: "It has been said that the two-letter word *no* is the single most effective time management tool there is" (tip No. 83). Then again, if you do master the art of saying no, you had better hope your friends haven't mastered time-saving tip No. 338: "When you meet a negative person, find a way to end the conversation and walk away. A negative person is one of the biggest wastes of time!"

Let's get this straight. Time-saving means getting enough sleep: "Don't kid yourself that by cutting back on sleep, you can get more done." Then again, time-saving means carving out "hidden time" for yourself late at night or early in the morning "with the help of a reliable alarm clock."

While awake, you should spend time daydreaming—"daydreaming time is quality time." But you should also be ready with a notebook to write down the ideas that come to you while you daydream.

You should take time to read poems and listen to music, especially adagios—having organized your CD's alphabetically by composer for quick retrieval. Yet whatever you do, don't procrastinate, and you might consider saving time by getting to the office thirty minutes before everyone else.

Once there, according to tip No. 209, you should stay off the phone by delegating 99 percent of your calls to your secretary. But according to tip No. 66, you should not only take every telephone call yourself but also have your secretary interrupt you with the next call. That way you can avoid piles of messages. Having a secretary in the first place is, of course, one of the most time-honored of all time-saving tips, but it doesn't help if you can't afford a secretary or if you are a secretary.

Time-saving is the subject of scores of how-to books published every year. They advertise their hopeful possibilities with titles like *Streamlining Your Life; Take Your Time; How to Have a 48-Hour Day; More Hours in My Day: Updated for the 90s;* and, shamelessly, *More Time for Sex: The Organizing Guide for Busy Couples.* The genre sells. Time-saving tips have become a Madison Avenue standby, as in, "Time Saving Tip No. 4: Don't take your kids to the store (Click here to buy groceries at NetGrocer)." It is easy to forget how very new in human history is the whole notion of time-saving. Personal time management did not exist as a distinct category in book publishing until the 1980s. The rare time-management titles of the last century, typically published by religious groups, advised readers on worthy ways to *spend* time, not ways to save time. Our culture has been transformed from one with time to fill and time to spare to one that views time as a thing to guard, hoard, and protect.

The experts who write these books reveal confusion about what it means to *save time.* They flip back and forth between advertising a faster and a slower life. They offer *more* time, in their titles and blurbs, but they are surely not proposing to extend the 1,440-minute day, so by "more" do they mean fuller or freer time? Is time saved when we manage to leave it empty, or when we stuff it with multiple activities, useful or pleasant? Does time-saving mean getting more done? If so, does daydreaming save time or waste it? What about talking on a cellular phone at the beach? Is time saved when we seize it away from a low-satisfaction activity, like ironing clothes, and turn it over to a high-satisfaction activity, like listening to music? What if we do both at once? If you can choose between a thirty-minute train ride, during which you can read, and a twenty-minute drive, during which you cannot, does the drive save ten minutes? Does it make sense to say that it saves ten minutes from your travel budget while removing ten minutes from your reading budget? What if you can listen to that audiotape after all? Are you saving time, or employing time that you have saved elsewhere, if you learn "how to have a 48-hour day" or "how to get 65 minutes out of every hour"?

These questions have no answer. They depend on a concept that is ill-formed: the very idea of time-saving. The first dictionaries to recognize *time-saving* as a word, barely a century ago, defined it as "prompt" or "expeditious" or "expedient." In a slow world, a time-saving device made an unpleasant task—washing clothes, perhaps—pass faster. Now we live in a faster world. Our time has different layers. It might seem that to *save* time means to preserve it, spare it, free it from some activity that might otherwise have consumed it in the hot flames of busy-ness. Yet time-saving books are constantly admonishing people to do things. Some of the recommended time-savers replace pleasant pastimes with less pleasant, for minutes or seconds. Some spare us a chore that was passing almost unnoticed in the background of our lives and replace it with a task that grabs more of our foreground attention. Saving time is a complex mission. Some of us say we want to save time when really we just want to *do more.* To leave time free, it is necessary to decide . . . to leave time free. It might be simplest to recognize that there is time—however much time—and we make choices about how to spend it, how to spare it, how to use it, and how to fill it.

The Telephone Lottery

By the way, tip No. 172 was: "The toll-free 800 telephone number listed on your computer software can save you time. If you don't find the answer quickly in your manual, don't waste time in a futile search. Call the experts."

Sure, try it. Hello, Microsoft?

The authors of a series of software-instruction books feel themselves to be working on a kind of time-loss frontier. Their very livelihood grows from the rising frustration with telephone technical support. The hours on hold, or describing symptoms, reporting the memory addresses implicated in page faults or system faults: this is "burn time" and "the bleeding-edge time sink, ours and yours." A typical problem takes them thirteen hours to resolve. Their E-mail is blocked—the E-mail supposedly saving many minutes that would be required to type, fold, seal, stamp, and post actual letters. But the authors are not saving minutes now. "It's our fervent belief," they write, "that this type of problem occurs every day and afflicts hundreds of thousands if not millions of people—both in business and personal affairs—simply trying to use electronic mail as a communications medium." The hours are spent in registering for support, finding product identification numbers, educating ill-prepared support staff, and then in the industry's notorious cure-all: removing and reinstalling software in toto. That approach, whether successful or not, consumes your time while allowing them to move on. There is little recourse. Scott Klippel of Austin, Texas, sued a software company in local small claims court in 1996 for lost time, and received what the company called a "nuisance" settlement, but most companies will not settle, and all are careful to obtain your implicit consent (ignorance being no excuse) to a licensing agreement disclaiming any "lost time" damages. We do sometimes lose our sense of proportion. The struggle, the trial, the chase can turn so many of us into little Captain Ahabs. "Once it becomes me against the machine, I don't like to admit defeat," Klippel said. "Even after the case was settled, I still kept trying to load the damned thing."

"Don't waste time" indeed. When you dial for technical support, you may not even get through. The telephone has created, along with many forms of supposed time-saving, one of the most peculiar and misunderstood forms of time wastage. The software industry alone leaves Americans waiting on hold for an estimated three billion minutes a year. Then there are the computer-hardware manufacturers, the airlines, the utility companies, the telephone companies themselves, and an incalculable number of government agencies. Like Dante's hell, the state of being On Hold has different levels. And before you get on hold, you must get past the busy signal.

You awake on, let us say, the last Monday in February. The

stroke of 7 A.M. Eastern time brings an annual ritual: across the United States, a few thousand people roll out of bed and simultaneously dial the Woods Hole, Martha's Vineyard and Nantucket Steamship Authority in hopes of reserving ferry tickets for their summer vacation. All but a lucky few will get a busy signal. That's what the redial button is for. Beginning two hours later, at 6 A.M. Pacific time, a horde trying to act on time-saving tip No. 172 will dial Redmond, Washington, and leap into the daily crucible of one of the world's most elaborate telephone queuing systems, the one dispensing Microsoft technical support. They will pay for the calls; no toll-free numbers here. Many callers will hang on long enough to reach an engineer; many will eventually give up. Then, all day, residents of New York and other big cities, seeking government services, will dial hundreds of equally notorious numbers, from the Brooklyn Post Office to the City Housing Authority to the Food and Hunger Hotline. And dial again, and again.

This is what might be called the Telephone Lottery—a kind of electronic triage. As a feature of life in the information economy, it has crept up on us. We take telephone lotteries for granted. Yet it might strike a visiting anthropologist from Arcturus as bizarre that, when the New York Yankees have one last batch of World Series tickets to sell through TicketMaster, they use the randomizing technology of the telephone network to sort winners from losers. How did the millisecond electronic decision-making inside the telephone switch become a nationwide arbiter of resources?

The ancient tradition of First Come, First Served has collided with the world of high-speed communication—a huge population of well-connected consumers making their demands felt in the same instant. In this sense the telephone network is the technological heir to a tradition that began when butchers and bakers grew large enough that they needed a "take a number" tape dispenser to queue up their customers. A store crosses one threshold of scale when jostling customers can no longer sort themselves out. Several more thresholds have to be crossed before a national clientele finds itself in an electronic queue organized by telephone wires. There is something perversely democratic about it, each citizen equally liable to place dozens of calls when, ideally, one call would do. In the name of fairness, telephone lotteries can cause network congestion on a national scale, as when desperation over Barbra Streisand tickets in 1994 engendered millions of nearly simultaneous calls. But the long-distance companies are not gravely concerned. "Of course, as a loyal Bell head I am deeply, deeply insulted when people use our network as a roulette wheel," said Greg Blonder, then director of customer expectations research for AT&T Laboratories. "But no more insulted, I suppose, than a toy store used by manufacturers to meter out Tickle Me Elmo's." In fact the telephone companies profit from telephone lotteries, at least in a small way: lots of people making lots of phone calls. Luckily for consumers, there is no charge for listening to a busy signal. The process is certainly more user-friendly than the overnight camp-out at the stadium ticket window or the line at a Moscow vegetable market. Still, in a world of economic rationality, might there be a better way?

The hidden costs come in the overuse of the telephone network. As Greg Sidak, a lawyer and economist with the American Enterprise Institute, notes: "That's a big externality. If the Yankees wanted to make it seem like they were being impartial or random in fulfilling orders for the tickets, they'd have to cook up some other technique—Ping-Pong balls in a fishbowl or something like that. Instead, they've shifted the costs onto the phone network and all the people who use it." The telephone network is big, but it is not infinite. It cannot handle unlimited numbers of calls, and in fact the circuits are designed with specific "blocking probabilities"—the likelihood of failure due to congestion.

Anyway, you're not getting those ferry reservations. The Steamship Authority sees its telephone lottery as a new and improved system. In the past, people made reservations by mail, which meant that thousands of letters needed to be sorted by postmark and processed in a single week. The authority's managers knew that did not happen. Now they have arranged matters so they will *not* know how many calls their customers make. The telephone lottery shields institutions from these delays. It transfers the loss of time to their customers. Too many letters means people working overtime and mail left unopened. Too many calls just means . . . busy signals.

So modern telephones come with a button that could hardly have been imagined in the early days of telephony: the button that redials the same number again and again. If your finger gets tired, you can try even newer technology: highly automated redialers like the PowerDialer, which can place your call up to twenty-five times a minute. It does this by taking the time-saving of Touch Tone telephones to the limit. The device uses a microchip to calculate just how fast the local switch will accept Touch Tone signals. Then again, advanced technology at the other end may push back, answering your call and placing you in a queue. If you call Microsoft, you can wait an hour or more, listening to music and occasional bulletins from a live "queue jockey," all the while paying your long-distance carrier, which has thus become an accidental beneficiary of crumbs from the Microsoft table. Other personal-computer companies have tightened the telephone choke-point to an extreme that led customers to file a class action. One client, a plaintiff's attorney charged, had assigned a secretary to do nothing but sit and call Leading Edge, the manufacturer of his troubled computer. Linda Glenicki, Microsoft's general manager of technical support, watches twenty or so of her company's scores of telephone queues, handling twenty thousand calls a day, in real time, on her computer screen. She can see the worst delays—eight minutes, eleven minutes, twenty-six minutes—but she maintains adamantly that, contrary to vast stores of anecdotal evidence, the company-wide average is barely a minute and a half. Why would anyone even complain?

She knows about telephone lotteries, though. She belongs to a hiking club, and when the time comes to reserve a spot on the popular trails on Mount Rainier, everyone calls at once. If you want to hike, you need a redial button.

Time Is Not Money

When Benjamin Franklin said that time is money, he wasn't just waxing poetic. He expected you to count it up:

> He that can earn ten shillings a day by his labours and goes abroad, or sits idle one half of that day, though he spends but six-pense during his diversion or idleness, ought not to reckon that the only expense; he has really spent, or rather thrown away, five shillings besides.

And Franklin never sat waiting in his car for the opportunity to throw a shilling into the exact-change basket at the George Washington Bridge.

Tollbooths are monuments of civic ineptitude—along with the telephone lotteries at city agencies and queues at unemployment and passport offices. Governments find it all too easy to tax the time of their citizenry. Arrive for your 4 P.M. appointment at the famous Bellevue Hospital, where a New York City doctor will examine your qualification for a Parking Permit for People with Disabilities, and you will discover that everyone—all the dozens of people who fill the waiting room—has a 4 P.M. appointment, so you will wait three hours for five minutes with the doctor. Long ago they tried giving people appointments at *different* times, but occasionally people were late and the doctor sat for several minutes with nothing to do. This is New York, where of all places on earth people feel most strongly the urge to keep moving, to shake a leg, to lose not a minute, but you realize that here at Bellevue they don't think time is money, or at least not a patient's time. True, you could hire a professional queuer to stand in line for you—these specialists can be found most often at Motor Vehicles Department offices. That implies a certain fungibility for time. But no accountants or auditors measure your lost time here or the billions of minutes expended in the cars funneling toward gatekeepers at the Holland Tunnel and the Bay Bridge and their thousands of traffic-choked counterparts worldwide. The toll collectors extract your ten minutes as ruthlessly as they do your two dollars. Only the dollars are counted. And only the dollars can be banked and spent. The time seized by governments evaporates.

As we have more to do and thus, by simple arithmetic, less time to do it, we have had to wonder more seriously whether and how time *is* money. Of course, we know it is. We know it from the economics of work: we are almost certainly paid by the hour. We pay our psychiatrists by the hour, even if it is a fifty-minute hour. We pay the telephone company by the minute. Television sells advertising by the second. Taxi meters formalize the connection. They are so engrained in our modern way of thinking that when we hear *the meter is running* we do not have to ask what meter; it is not energy or water trickling away. Automated teller machines formalize the connection, too, in their own way: "All-night banking," observes Mark O'Donnell. "It never stops. Go go go. Go Go Boy. Satan never sleeps." Our lives are filled with border crossings where we need to make a currency exchange—trade dollars and cents for hours and minutes if we can. We make this trade at the airport gate when the airline—having overbooked a flight in the service of maximum efficiency—holds a concealed reverse auction and offers cash to travelers willing to be bumped. Economists struggle to take the arithmetic further toward its logical conclusion, whatever that might be. They consider the "short-run allocation of discretionary time." They consider the "scarcity value of time"—wouldn't it be appropriate to equate that with the marginal wage rate? Henry Ford got his two cents in early: "If a device would save in time just 10 percent, or increase results 10 percent, then its absence is always a 10 percent tax. If the time of a person is worth fifty cents an hour, a 10 percent saving is worth five cents an hour." He believed in time-money arithmetic, in other words. So did the authors of a Brookings Institution study on the costs and benefits of automobile seat belts in 1987. They argued that previous researchers—who had found considerable value in the use of seat belts—had failed to consider the *time* spent in buckling and unbuckling. With weird precision, they calculated the "sample mean time to fasten belts" as 2.97 seconds. "Our seat belt use equation," they admitted, "does not facilitate a direct calculation of the trade-off between money and time." So they calculated it indirectly and produced a figure of $2,169 over the life of a car.

Yet even economists have moments of rationality when they recognize that the passing seconds of our lives cannot be rolled up and exchanged quite as easily as the pennies that accumulate in our pockets. If the efficiency expert from hell were to sit beside you with an infernal stopwatch and clock your seat-belt fastening at 2.97 seconds, could you really say that you would save 2.97 seconds by not fastening the belt? Is it time that you could use instead to moonlight at your marginal wage rate of $18 an hour or to listen calmly to an extra bar of a Beethoven piano sonata? All right, not really. The very existence of this arithmetic, the plausibility or near-plausibility of the research, stands as evidence of our burdensome relationship with the minutes and seconds. Ford actually meant his calculations to sound humane. "The idea is that a man must not be hurried in his work," he wrote. "He must have every second necessary but not a single unnecessary second." Ford himself, as the century's archetype of a factory owner, made sure he was the master of those seconds. Like all good metaphors, *Time is money* has a degree of truth that varies with where you stand. For Henry Ford, time was money and was not money; he was not paid by the hour. George Lakoff and Mark Johnson, in *Metaphors We Live By,* take pains to note that "time isn't really money":

> If you *spend your time* trying to do something and it doesn't work, you can't get your time back. There are no time banks. I can *give you a lot of time,* but you can't give me back the same time, though you can *give me back the same amount of time.*

This is the clever sort of wordplay you might expect from academic literary theorists with time on their hands. But it is also a conundrum likely to burden practical economists for the next generation or more. Time, not money, takes center stage in the new economy. We buy and sell computer time and golf time. The spirit of these transactions would have been abhorrent to the Middle Ages, when the Church understood that time belonged to no one but God and that trying to sell it amounted to usury. We are not so finical. Our modern economic life depends increasingly on the scarcity of time, the competition for time, the revaluing of time, and the redistribution of time. Where are the equations for that?

Broadcast networks, roadside restaurants, Internet marketers, even grocery stores and automobile dealers have all learned that their first objective is a slice of your day. They value *your* time. Smooth salesmanship still means persuading you to loosen your fingers, let go of that stuff you cling to so dearly, but those coins are minutes and seconds. "You give us twenty-two minutes and we give you the world"—deal? In this tightly knit world, where virtually anyone with a product or service can reach you, the competition for your *time*—never mind your money—grows fantastically intense. Here is Freeway, a service in Pittsburgh, offering free long-distance telephone time to anyone willing to listen to a quick commercial message. It's a simple trade: you listen for fifteen seconds, and then you can talk for exactly two minutes. You can just pretend to listen. But you must hand over those seconds. Does it feel demeaning? You do it for the television networks, and you will do it more and more. Maybe you will find a way of charging junk mailers for the privilege of putting unsolicited advertisements into your in-box. A whole new industry specializes in "on hold" advertising. "The caller you placed on hold is your captive audience," says a promotion from Ontherun.com. "You may have hundreds or even thousands of captive prospective customers placed on hold each year. This is the perfect time to promote your company." Everyone is trying to grab you by the lapels.

The same phenomenon means that merely delivering a product at a low price is not enough anymore. Marketers must engage in marketing. Merchants must provide excitement and novelty. At all costs they must not *bore* you. Horseradish mustard? You've been there and done that. How about roasted-garlic mustard? Yet the more novelty the economy offers its consumers, the more quickly they seem to grow jaded. And how, in a time-driven economy, are businesses supposed to handle the accounting? It is just as Lakoff and Johnson say, in their altogether different context: businesses get you to *spend* your time, watching *ABC News* or eating at Burger King. But they can't *bank* your time. It comes off your ledger, but they usually can't add it to theirs. Somehow they have to collect money too. And in seeking your time, they have to be careful not to grasp for too much.

Complex as the bonds between time and money already are, businesses add more, like the assistants tightening chains and handcuffs on an escape artist, only this time it seems there will be no escape. Fumio Komatsuzaki, manager of a Totenko restaurant in Tokyo, heard about a local pond charging anglers by the minute, and this gave him an idea. He installed a time-punch machine for his customers. He now offers all-you-can-eat—at a rate of thirty-five yen per minute. So the diners rush in, punch the clock, load their trays from the buffet table, and concentrate intensely on efficient chewing and swallowing, trying not to waste time talking to their companions before rushing back to punch out. This version of fast food is so popular that, as the restaurant prepares to open at lunchtime, Tokyo residents *wait in line*.

You settle for the McDonald's drive-through line. Drive-through hamburgers are routine, of course, in the time-scarce economy, where you can find drive-through cash and drive-through liquor. Something goes wrong, though, and you find yourself shunted aside, waiting hungrily in the parking lot for someone to deliver your order by hand. Minutes pass, and a tiny photocopied text arrives through your window. It appears to be a message from command central in the war against time wastage:

> About our Drive-Thru
>
> Question: Why do we say "Please move away (forward) from the window" or "Please have your money ready before coming to the pick-up window"?

> Answer: We want to serve you, and any potential customers after you, as fast as we can. We have installed an electronic sensor to monitor how long your car is at the pick-up window. If your car is waiting at the window for more than 40 seconds, a bell rings and a counter increases by one. This penalizes our performance score.

Performance score! But there wasn't even another car in line, you sputter, before reading on:

> We know sometimes there is no car behind you, but we ask you to move forward quickly anyway because the electronic sensor cannot detect this condition and will still penalize our performance score.

No wonder McDonald's, in less than a half-century of existence, has become one of the world's most powerful brands. Analysts attribute its success to familiarity, cleanliness, standardization, and low prices—almost forgetting the first attribute of *fast* food. The company does not forget. Its original mascot, before Ronald McDonald, was a hamburger-shaped character named Speedee. The company aims to have a restaurant within four minutes of every American. In 1997 its marketers tried to speed up even more by offering refunds to any customer not served within fifty-five seconds. The franchise-holders rebelled at this, but they do keep hiring hamburgerologists who train in briskness and efficiency at the company's Hamburger University, in Elk Grove Village, Illinois. The faculty applies the highest level of time-and-motion expertise to burger assembly and french-fry scheduling. The hamburgerologists are worthy of their enemies: the epicures of the international Slow Food Movement, founded at the Opéra Comique in Paris in 1989, where delegates chose the snail as their emblem, suitable for lapel pins, and endorsed this manifesto:

> We are enslaved by speed and have all succumbed to the same insidious virus: Fast Life, which disrupts our habits, pervades the privacy of our homes and forces us to eat Fast Foods. . . .
>
> May suitable doses of guaranteed sensual pleasure and slow, long-lasting enjoyment preserve us from the contagion of the multitude who mistake frenzy for efficiency.

Let us make no mistake about frenzy. By 1998, the Slow Food Movement had spawned a Virtual World Guide to Slow Places, on-line, where it promised "a multi-tier service with an archive consultable on the Internet and more personalized E-mail, telephone, or fax facilities for Slow Food members." Thank you, epicures! By all means, spare us the contagion of the multitude.

If time has become a commodity this precious, we naturally start to think in terms of budgets. We resent the time spent navigating that peculiar late-twentieth-century human-machine interface, the telephone voice-response system. We wonder whether to believe the "estimated wait time," we try to remember the "hot keys," we fear getting caught in never-ending loops, and all too often we "bail out," by pressing 0 or pretending we don't have Touch Tone buttons at all. Usually, though, we surrender. Sometimes these automated systems, handling hundreds of millions of calls yearly for a typical telephone company or electric utility, really do save us time; often, however, they actually cost the consumer time—they effect an involuntary transfer of time from our ledger to the company's. Is it really our choice, when the recorded voice holds out a promise of time-saving: "If you have a Touch Tone phone and would like to use our *express* service instead of *waiting* for a representative, press 1 now"?

Naturally we chart our leisure time long in advance. Summer vacation rentals in the popular resort areas of the northeastern United States are often gone by the previous autumn. Grand Canyon Lodge takes reservations two years in advance and sells out quickly; thirteen months in advance of Christmas dinner at Ahwanee Lodge, tens of thousands of people try to sign up. If you value your time highly enough, an ordinary low-key vacation may seem inappropriate—extravagant in a different way—so maybe you will hire hunting consultants to arrange a $19,000 Stone Sheep safari in the Yukon; if so, you will find the best areas spoken for two years in advance. Wedding halls and churches in populous cities are often booked two years ahead. Disney World reports that "consummate planners" make reservations five years ahead of time. Then they wait in line and hurry forward. We consumers act like runners sprinting by, with merchants of leisure time lining the roadway trying to get us to glance back over our shoulders.

We know that money can be acquired, saved, and spent. We sometimes act as though we could treat time the same way. But we don't save it. We shift it to different activities; or we use it; or we simply live. By tradition, impatience is a vice. Haste makes waste. Even if our technological world seems inspired by the modernist calculations of Benjamin Franklin, we can all think of a few remaining human activities that cannot profitably be rushed. "There are two cardinal sins," Kafka said, "from which all the others spring: impatience and laziness." There's the paradox—maybe it's laziness, not industriousness, when we succumb to the economics of time.

Short-Term Memory

As the flow of information accelerates, we may have trouble keeping track of it all. In past times companies stored data on punch cards, as rows of holes; then on big, soft, eight-inch floppy disks, or on magnetic tapes, like Univac's Type II-A. These, unfortunately, grow ragged and faint over years of sitting in the ghostly magnetic fields that are part of life on earth. Maybe your company just saved data on mag cards for the mag-card IBM Selectric Typewriter. "And where could you get one of those today, and why would you want to?" said Ken Thibodeau, director of the Center for Electronic Records, responsible for the archiving of the uncountable records of the United States Government. IBM's published list of Discontinued Storage Media grew longer and longer: optical disks, data cartridges, mini-data cartridges, maxi-data cartridges, diskettes of all sizes, and somehow, mixed in with these, "Fifty file tab dividers." Eight-inch floppy disks had been a magnificent improvement over punch cards, and few were sorry to see *file clerk* begin to fade from the corporate vocabulary. But obsolescence came faster and faster. For that matter, film, meaning 16- or 35-millimeter acetate, fell victim to the far more convenient but otherwise inferior technology of videotape. Old prints of great films, along with people's home movies, faded, burned, moldered, or just got lost. The life cycles of storage media for the data coursing through computers became as short as two to five years. We now stockpile our heritage on millions of hard drives and optical disks, and these flaky objects, too, promise to go obsolete on a rapid schedule.

Many of the world's librarians, archivists, and Internet experts see a crisis looming. They warn that our burgeoning digital culture is heading for oblivion, and fast. "There has never been a time of such drastic and irretrievable information loss," says Stewart Brand, creator of the *Whole Earth Catalog* a generation ago. Our collective memory is already beginning to fade away, he argues. Future anthropologists will find our pottery but not our E-mail. "We've turned into a total amnesiac," Brand says. "We do short-term memory, period." The information-storage medium of the past couple of millennia—for words not writ in stone, anyway—has of course been paper. Paper does decay with time, and it is fragile. One fire at the library at Alexandria in 391 C.E. destroyed a big piece of the ancient world's heritage. But to some people, paper is beginning to look good. As consumers of technology we're easily seduced. We mothball three-year-old PC's. But the data have time scales of their own, perhaps measured in centuries. Some companies have begun "refreshing" their aging records, by continually copying them onto new storage media using new software. Refreshing isn't easy, and most institutions have not yet realized that it may be necessary. Whatever media they use to save their digital information, they will not be able to read it without a machine—a finicky antique, most likely. With paper, all you need is your eyes.

Perhaps the speed and richness of the Internet have lulled us, letting children in Boise read census data from Washington and oral history from Hiroshima. Words swim instantly across the network, not caring about the mileage, and we don't exactly feel information-deprived. We may be drowning, actually. But are we sacrificing longevity to gain glut?

It's scary. And yet . . .

Anyone wandering through the Internet might begin to feel that memory loss isn't the problem. Archivists are everywhere, in fact—official and self-made. The leading on-line bridge service has recorded every detail of the bidding and card play in each of the millions of hands played since the early 1990s. Likewise, any silly message that you broadcast to a Usenet newsgroup is now being stored, for eternity or some approximation thereof, by a variety of commercial services. No matter that you gave your last posting a mere five seconds' thought; you should be prepared to hear your biographer read it back to you in your dotage. Most people, unfortunately, don't have posterity in mind when they fire off their little notes. Internet communication seems so spontaneous and personal. Will people really want future employers to dig up all the messages they've been posting to alt.dead.porn.stars and soc.support.depression.manic? Sometimes, as the years go by, privacy demands a gentle forgetfulness.

Many people sitting at company workstations toss off their E-mail as casually as they speak—gossipy E-mail, secretive E-mail, snide E-mail, raunchy E-mail, E-mail meant to vaporize after serving its instant purpose. But it does not disappear, as corporate lawyers across the country have realized. Neither sender nor recipient can delete it reliably. To the lawyers' occasional horror—here comes the subpoena!—it lingers on disk drives and backup tapes like a late-night guest who has forgotten how to leave.

The biggest proprietor of archivable data is the federal government, struggling to preserve the records it generates daily on an uncountable scale. Literally uncountable: the last serious attempt was made early in the 1990s by the National Academy of Public Administration, which found—excluding the vast stockpiles of scientific data at the space and weather agencies, and data on individual PC's—about twelve thousand major databases; and the researchers also estimated that they had probably missed about the same number. Public interest groups sued to ensure that every piece of governmental E-mail be preserved as a "federal record." Either way, the task of the National Archives and Records Administration is monumental. "Digital information technology is creating major and serious challenges for how we're going to preserve anything of our culture and our history," says Thibodeau. "It's also creating opportunities: we'll be able to preserve and use a lot more information than ever before." Pity the poor historian, though. The Clinton administration's E-mail for the Executive Office of the President alone exceeded eight million files.

Meanwhile, in its unofficial way, the Internet is transforming the way information is stored. The traditional function of libraries, gathering books for permanent storage or one-at-a-time lending, has been thoroughly confused. Archiving of the on-line world is not centralized. The network distributes memory. There is a kind of self-replication at work, with data employing humans in the effort to spread and reproduce. Web site by Web site, the data seem as frail as skywriting—smoke in the breeze. Brewster Kahle, estimating the average lifetime of a Web page at seventy-five days, created an Internet Archive to capture and store periodic snapshots of almost the entire Web. It saves pages that have been lost or shut down by their owners. It amounts to about eight terabytes of data. (*Tera-* is trillion; *peta-* is next.)

Archivists have new practical problems to struggle with. Who, if anyone, will decide which parts of our culture are worth preserving for the hypothetical archaeologists of the future? Can any identification scheme help readers distinguish true copies from false copies in the on-line world's hall of mirrors? What arrays of optical or magnetic disks might provide reliability and redundancy for more than a few years of storage? Still, hope comes from the simple truth that the essence of information does not lie in any technology, new or old. It's just bits, after all.

In the world before cyberspace, countless bridge hands were played and words spoken, and the memory vanished like vapor into the air. All that information, dissolved no sooner than it was formed. Once in a while people managed to snatch a bit back from the ether, with pen on paper or, later, audio- and videotape. They succeeded in saving for posterity a tiny portion of what was worth saving: the speeches of Lincoln (the major ones), the poetry of Shakespeare (but not quite reliably), the plays of Sophocles (except the lost ones), and a few dozen terabytes more. Once it was expensive and slow to capture a visual image and preserve it for the eons. If you were a successful Dutch merchant of the seventeenth century, perhaps you could afford to hire Rembrandt Harmenszoon van Rijn to knock off a quick portrait. Then another trail of exponentially accelerating technology: letterpress engravings on zinc, wet collodion photography, studio photography, halftones, Muybridge, Kodaks, Polaroids, eight-millimeter home movies, handheld videocameras, ten-dollar disposable cameras available at drug stores, Internet Web cams, traffic cams, office cams, beach cams— All those photons used to scatter into the void. Now we trap them for recycling.

We *know* the world is changing fast; we know we are nearsighted; we berate ourselves for our foreshortened time horizons, and we bury our detritus as lovingly as dogs burying bones. We bury it in time capsules, for example. The business of time capsules—once a rare bit of whimsy at world's fairs—has grown into an industry. The International Time Capsule Society estimates that more than ten thousand people and institutions have buried time capsules. They must think that future archaeologists will be grateful for the bounty of twentieth-century wristwatches, telephone books, decorative caps, CD-ROM's, and ampules of Budweiser beer. "Attention: Schools, Colleges, Corporations,

Businesses, and the Rest of Humanity," screams an advertisement from A-1 Time Capsules of Bradbury, California—A-1 offers noncorrosive plastic cylinders with optional metal plaques. Other companies include a spray of inert gas, no extra charge. One town council wanted to deposit some videotapes; its consultant, Greg Blonder, tried to explain that the tapes would "rust" from magnetic domain reversals and become useless as VCR's inevitably went obsolete. "They couldn't believe there would be a time without videotapes, despite millennia of experience without even a TV," says Blonder. "And when we showed them how sulfur compounds outgassing from the 1993 championship football would cause all the paper in the capsule to yellow and crack, things got a little tense."

The future packaging industry, as it calls itself, depends on the peculiar misconception that the future's problem will be not having enough of *us*. Future packagers recommend visiting museums to get ideas about what might appeal to future curators—as though museums themselves were not multiplying feverishly. They suggest clipping end-of-year wrap-ups from newspapers and newsmagazines, not noticing, apparently, that those are already being saved elsewhere. Most of all, the Internet turns a large fraction of humanity into a sort of giant organism—an intermittently connected information-gathering creature. Really, amnesia doesn't seem to be its worst problem. This new being just can't throw anything away. It is obsessive. It has forgotten that some baggage is better left behind. *Homo sapiens* has become a packrat.

Would it be overreaching to say, "There is no practical obstacle whatever now to the creation of an efficient index to all human knowledge, ideas and achievements, to the creation, that is, of a complete planetary memory for all mankind"? Those are H.G. Wells's words, written in 1937. "And not simply an index," he continued. "The direct reproduction of the thing itself can be summoned to any properly prepared spot."

"This in itself is a fact of tremendous significance," Wells wrote.

> It foreshadows a real intellectual unification of our race. The whole human memory can be, and probably in a short time will be, made accessible to every individual. And what is also of very great importance in this uncertain world where destruction becomes continually more frequent and unpredictable, is this, that . . . it need not be concentrated in any one single place. It need not be vulnerable as a human head or a human heart is vulnerable. It can be reproduced exactly and fully, in Peru, China, Iceland, Central Africa, or wherever else. . . . It can have at once, the concentration of a craniate animal and the diffused vitality of an amoeba.

Wells was not imagining the internetworking of computers, of course. The new information-storing technology that inspired him was microfilm. He had no idea how fast it would go obsolete.

The Law of Small Numbers

In 1876, when people thought the world was beginning to grow pretty large, Colorado was admitted to American statehood and called itself the Centennial State. Back east, the centennial was also celebrated with the first international trade fair, the Philadelphia Centennial Exposition. Thirteen years later, the French organized a Centennial Exposition and built Gustave Eiffel's tower to commemorate their own revolution. Four years after that, in 1893, a Chicago fair celebrated the fourth centennial of Christopher Columbus's discovery of America (no one had thought to mark any of the first three centennials). In 1939 a crowd in Cooperstown, New York, dedicated a new Hall of Fame and declared a somewhat fictional centennial of the invention of baseball. In 1976, United States football celebrated the centennial of its playing rules.

The centennial impulse has ancient roots. It draws on our love of round numbers, our enjoyment of celebrating, and perhaps our slightly wishful view of the human life span. In 1617 Protestants across Europe marked the hundredth anniversary of Martin Luther's posting of the Ninety-five Theses. The centennial is an odd creature nonetheless. It is self-referential, hermeneutic, an excuse for parties, a "pseudo-event" in the sense of Daniel Boorstin—an occasion that exists only for the sake of publicity. A pseudo-event does not actually happen, and in the course of not happening it can consume considerable money and public attention. Oddly enough, in our accelerating and crowded age, the centennial may be headed for a collapse under its own weight. The year 2000, a very round number, will engender more than a few centennials. No doubt the proprietors of the *Guide Michelin* will observe the hundredth anniversary of its birth. So will the Métro in Paris. So will the National Automobile Show, the Davis Cup in tennis, the American League in baseball, and the International Ladies Garment Workers Union. The multiplication of centennials outpaces the multiplication of actual events, because people and institutions typically provide not one but many choices of suitable dates, beginning with birth and death. Concert halls and theaters in 2000 will surely observe the centennials of Aaron Copland (born 1900), Oscar Wilde (died), Kurt Weill (born), Arthur Sullivan (died), and Louis Armstrong (born). Kodak will have occasion to remind its customers about the birth of the Brownie hand-held camera. Psychiatric associations may note the centennial of *The Interpretation of Dreams.* Scientists could plausibly choose 2000 as the year to mark the hundredth anniversary of many different achievements, from blood typing to quantum physics. The world really is growing larger, and the effects are peculiar.

Richard K. Guy, a mathematician in Alberta, Canada, asks you to think about sequences of numbers. Even if we're nonmath we make room for some of these in our heads. Any New Yorker, for example, will recognize:

- 14, 18, 23, 28, 34, 42, 50, 59, 66, 72, . . .

There are people who spend their lives analyzing and cataloguing these sequences—tens of thousands of them. Some are ordinary. Some are sublime. Some are ridiculous.

- 1, 4, 9, 16, 25, . . . These are the square numbers (1×1, 2×2, and so on).
- 1, 2, 4, 8, 16, . . . The powers of two, of course.
- 1, 20, 400, 8902, 197281, . . . The number of possible chess games *n* moves long (0 moves, 1 move, 2 moves, and so on).
- 1, 4, 11, 16, 24, . . . Aronson's Sequence, defined by: *T is the first, fourth, eleventh, sixteenth, twenty-fourth . . . letter in this sentence.*

The second sequence is also the number of layers in a piece of paper folded in half *n* times. That is not a coincidence. The sequence also happens to represent the number of regions into which a circle is divided by lines connecting *n* points spread around its rim.

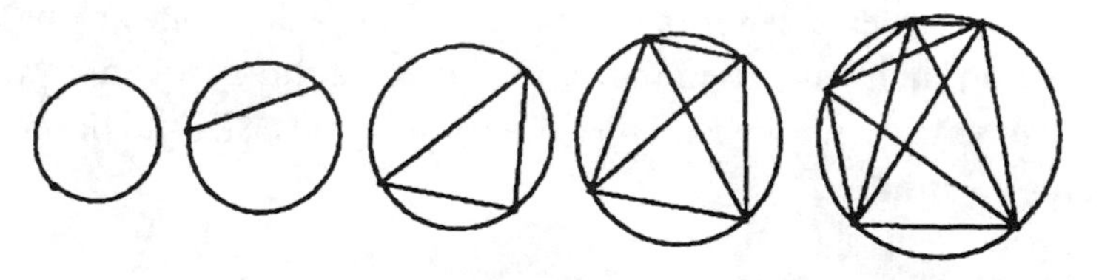

That, however, *is* a coincidence, and the coincidence does not last. The powers of two continue in their merry way (. . . 32, 64, 128 . . .), while the circle-slicing sequence goes off on its own

(. . . 31, 57, 99 . . .). If you looked only at the small numbers in the sequence, you might be fooled about the pertinent mathematical rule.

The mathematicians who track these things have noticed how often the small-number sequences appear to be doing double duty, or worse. Here is another one: 1, 1, 2, 5, 14, . . . This could be the number of different ways of folding a strip of *n* postage stamps into a little stack.

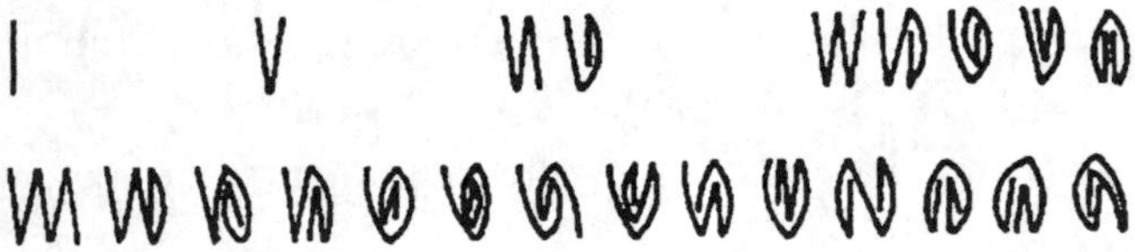

Or it could be the number of distributions of *n* distinguishable objects in indistinguishable boxes, with at most three objects in a box.

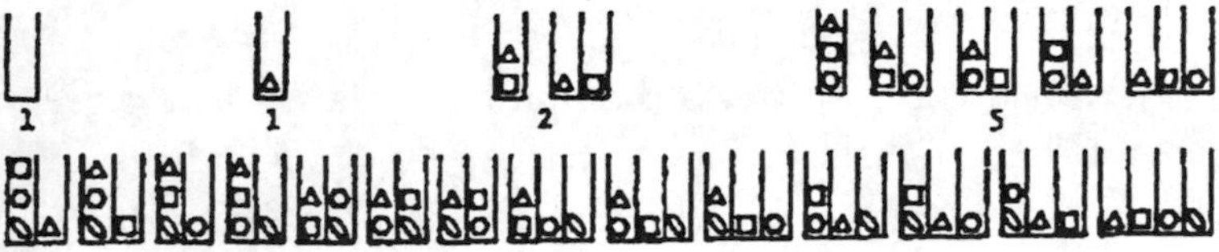

Or it could be the number of different groups, up to isomorphism, of order 2^n. (Sorry, no picture.) You would have to be a very sharp mathematician to know that these three sequences are fundamentally different despite their identical start. The postage stamps continue:

- 1, 1, 2, 5, 14, 39, 120, 358, 1176, 3527, 11622, . . .

The distributions:

- 1, 1, 2, 5, 14, 42, 132, 429, 1430, 4862, 16796, . . .

As for the groups of order 2^n:

- 1, 1, 2, 5, 14, 51, 267, and after that no one is frankly quite certain.

So that innocent-looking sequence 1, 1, 2, 5, 14 bears a heavy burden.

A growing collection of observations of this kind led Guy to formulate and "prove by intimidation" what he calls the Strong Law of Small Numbers. It is not a purely mathematical observation. To put it simply: "There aren't enough small numbers to meet the many demands made of them."

This law is the enemy of mathematical discovery, Guy says. A mathematician sees a pattern. Sometimes the pattern persists forever. Sometimes the pattern is a figment, and disappears when we reach the realm of large numbers. In focusing on sequences of numbers, mathematicians are studying, analyzing, and classifying the ways in which the purest tendrils of things unfold as they go from small to large. These are patterns made of logic. Anyway, as a rule, the realm of small numbers is misleading. One-fourth of the first one hundred numbers are prime numbers; one-tenth are perfect squares. These nice things quickly get rarer among the large numbers. For that matter, if you looked only at the small numbers, you would think numbers were very likely to be Fibonacci numbers, Bell numbers, Catalan numbers, Motzkin numbers, and even perfect numbers. Of course, you would be wrong. No less a master than Pierre de Fermat looked at the first numbers in the sequence of powers of powers of two plus one ($2^{2^n} + 1$: 3, 5, 17, 257, 65537, . . .) and determined that they were, and all their successors would be, prime. He was wrong. Many errors in the same family have followed along behind.

Is there a message for us here? Back in the real world, in a simpler time, the Columbia Broadcasting System, the German Nazi party, and the Daimler-Benz and Chrysler car companies chose iconic forms for their company logos. Now marketing specialists have more trouble finding simple, memorable, geometric shapes suitable for logos; logo creation has become a multimillion-dollar business. You could probably identify the Chrysler icon, the pentagon with five equiangular spokes, even out of context. The Mercedes icon, the circle with three spokes, has also been spoken for. For a car maker hoping to lay claim to a memorable little icon to be milled in metal and stuck on hoods, how many possibilities are there? A small number.

Also at mid-century, the number of varieties of prepared mustard available within ten miles of the average resident of the industrial world—a number that stood at zero, of course, through most of human history—was rising toward one. Now, in the most ordinary condiment aisle, after Gulden's and Dijon come English, Bavarian, ball-park, Habañero, dipping, island, Creole, horseradish, brown, tarragon, honey, Indian, classic, three-alarm, purple—no, we will not go there. Let's just say the number was formerly small and now is large. Thus choosing the proper mustard takes time, not to mention savoring it.

To name a new medicine once required a few moments' thought. Now a pharmaceutical company knows that it will conduct a proprietary name evaluation as part of its labeling review at the Food and Drug Administration. It will be prudent to request an early consultation with the FDA's Labeling and Nomenclature Committee, never forgetting the Patent and Trademark Office and the United States Adopted Names Council. The domain of drug names is densely overcrowded, and the density carries particular dangers because confusion can be deadly. Is Rezulin the new insulin enhancer or the old anti-acne medication? Is Dynacin the antibiotic and Dynacirc the antihypertensive, or is it the other way around? The package designers may wish to use the letters NS for their nasal spray; will they know that doctors also use them as shorthand for "normal saline"? The process of inventing a drug name routinely takes many months' work by expensive consultants—Brand Institute, Name Lab, Lexicon—even before considering cross-language difficulties. Another sign of this overcrowded name space is the tendency to capitalize brand names in the middle: market a high-speed network-access method and call it PeRKInet, as though we now had a fifty-two-letter alphabet. Anyone who has tried to find a fresh or unique name for a brand, an Internet domain, a children's book, a rock band, a space vehicle, or a perfume has stepped into a packed, finite space and bumped into the Strong Law of Small Numbers.

This intensity, this swarming, comes with our greater reach. Our choice of shoes is global; cobbler, farewell. Just as computers make it possible to see larger Mersenne primes than Fermat imagined, they make it possible to link the world's number theorists and would-be number theorists. You could join the Great Internet Mersenne Prime Search and subscribe to the Mersenne Prime Mailing List, so that hours would not pass before you would learn of the discovery in 1997 of the thirty-sixth, an 895,932-digit number, by a man in England using a computer program written in Florida, or the thirty-seventh, a number that would fill an even thicker book, the next year. Marin Mersenne himself made quite a few errors, it turns out. Then again, he lived in a cloister. You do not; or, if you do, your cloister has its own Web site. When you stretch out an arm to buy mustard, no delicatessen on earth is too far. When one hundred years roll around, you have a plenitude to celebrate.

The Strong Law of Small Numbers tells us something about the increasing complexity that so often triggers that sense of hurriedness. Like the small numbers, the words of two syllables and the basic condiments and the central television networks bear a heavy burden. They are placed under strain by access to the varied words and tastes and video programming sources that lie beyond. All our information sources evolve toward complexity. No software program gets simpler in release 2.01. No television-news anchor or daily newspaper holds its former central position as announcer to a whole nation. Instead citizens awaken each day with a multitude of experiences to divide one from the other—last night's five hundred channels and million Web sites. Yet these complex strands sometimes return to a simple point of origin. The focal points of national obsession become, if anything, more furious and intense: the trial of O. J. Simpson, the perils of Monica Lewinsky, the coming of—dare we say it—the millennium. Andy Warhol is less famous for any art he may have left behind than for his observation, "In the future, everyone will be famous for fifteen minutes." The World Brain's attention span may seem short. Its ability to focus on any one celebrity may seem to have waned, but that is because the pageant flitting before its eyes is so crowded and multifarious, not because fame is so easily had. The ranks of the unfamous and invisible have also swelled. Woody Allen's gloss is correct: "Almost nobody will be famous for even one minute." The connections between complexity and speed—between variety and time pressure—are not always obvious, but they are real.

In 1996 the American trademark authorities noted 234 applications for different products to be named with the word *millennium.* The next year the number rose to 404. By early 1998 the pace of applications had more than doubled again. Companies attempted to reserve for themselves the label of "official" airline, candy, hole-in-one prize company, light bulb, water, vending machine, baby, souvenir, retirement planner, Champagne, public-relations firm, vacation, Web site, and sponsor of rapid tooling and prototyping stereolithography, of, in, and for the millennium. They asked to trademark Millennium sewing machines, mutual funds, air fresheners, popcorn, metal, magic, fluids, collectibles, great moments, bathing systems, bottling services, golf tournaments, moon monsters, heroes, dynamics, coins, bombs, minutes, law firms, injury law firms, and personal injury law firms. They grabbed and invented spellings for Billenium, Hillenium, Malenium, Pharmillenium, Mealleaniyumm, Millenion,

Millenifix, Milleniatron, and Mil-Looney-Um. Slogans already spoken for include Have a Nice Millennium, New Millennium Madness, Navigating the New Millennium, Rock the Millennium, Working Straight Through the Millennium, We Survived the Second Millennium, This Is the Millennium, and Only One Company Offers So Much on This Side of the Millennium.

Then there is the land rush around the terms Y2K and twenty-first century and the number 2000. Did you hurry? Or did you wait till the last moment? Confronted with complexity, our instincts seek order, pattern, simplicity. We humans are geniuses at distillation—we automatically take the buzzing, teeming richness of experience and find a manageable set of objects or laws. Or is that set manageable after all? The Strong Law of Small Numbers says it tends to get overloaded. What could be more reassuring than a round number like 2K? Yet the millennial panic starts welling up . . .

We are small-number people in a large-number world.

Bored

Light is good. Yet in the dark the stars come out. You have to wait long enough for your eyes to adjust to the darkness. The human eye is an instrument capable of resolving a very large range of intensities, but not all at once. The very bright and the very faint cannot be seen together. Finally, accustomed to the dark, you see what was invisible in the light.

Speed is like light: something we crave, all else being equal, and something that can obscure as well as illuminate.

Someone puts on an obscure record. Maybe it is a motet from the early Renaissance, *Absalom, Fili Mi,* by Josquin des Prez. You cannot really enjoy this if you are living at the normal modern speed, although it takes only four minutes to get through not all that many notes. Unless you happen to be an aficionado of slow polyphony, it will sound nice and bore you. Josquin would not have appreciated the typical Nike commercial. There is a way, though, to perceive the anguish of this old sound bite, to descend with Josquin into the depths of hell. Try a few minutes of sensory deprivation first; let yourself get a little bored. *Then* put the record on.

You are bored doing nothing, so you go for a drive. You are bored just driving, so you turn on the radio. You are bored just driving and listening to the radio, so you make a call on the cellular phone. You realize that you are now driving, listening to the radio, and talking on the phone, and you are still bored. Then you reflect that it would be nice if you had time, occasionally, just to do nothing. Perhaps you have a kind of sense organ that can adjust to the slowness, after being blinded by the speed. The void is not so dark after all. With the phone not ringing, the television switched off, the computer rebooting, the newspaper out of reach, even the window shade down, you are alone with yourself. The neurons don't stop firing. Your thoughts come through like distant radio signals finding a hole in the static. Maybe they surprise you; maybe they disturb you; maybe they assemble themselves into longer strands—ideas, or knowledge, that might not have formed in the usual multitasking hurly-burly. That is the view, anyway, of some plangent advocates of leisure—Sebastian de Grazia, for example, who declared in 1962:

> Perhaps you can judge the inner health of a land by the capacity of its people to do nothing—to lie abed musing, to amble about aimlessly, to sit having a coffee—because whoever can do nothing, letting his thoughts go where they may, must be at peace with himself.

Yet he remembered, too, Aristotle's comment that the Spartans collapsed as a society when peace came. They had been competent and fulfilled in the business of war and utterly flummoxed by the different demands of leisure. In short, they were *bored*—though no such word existed in Greek, or in English until very near to modern times.

When the time comes to be alone with ourselves, we may crave the cellular phone at that. As nature abhors a vacuum, so we abhor the blankness, the lack of stimulation, that comes with *doing nothing.* Activities rush in to fill the void—and never have so many interesting activities been available. Maybe, alone with our thoughts, we feel that there just isn't enough to keep us entertained. Thoughts do have a tendency to ramble, jump around, repeat themselves, and otherwise fail to become radiant and monumental. We can try to organize our aloneness through prayer or meditation. We can try to focus long enough to build nontrivial coherent chains of thought. Mostly we need priests for our confessions or psychotherapists for our inbound adventures of discovery and healing. It turns out to be difficult to travel far without a guide, even through our own familiar selves. Writing, too, is a way to create, by accretion and continual reorganization, more than can be assembled merely alone with one's thoughts. And we are social animals. Language was not invented for improving the quality of introspection. People in pairs and people in groups can usually create more interestingly and entertain one another more richly than can our solitary selves. Must we feel guilty if we cannot be satisfied doing nothing—if we don't like to do nothing?

All those clamoring activities line up by rank, in order of the power of their claim on your attention. That book looks appealing, but this magazine pulls harder. Even better is that new jazz recording, but then you prefer the exhilarating rush of an on-line session of the game so fittingly called Total Annihilation. It's as if, corrupted by haute cuisine and soft mattresses, we can't go back to the simple pleasures of plain bread and butter and sleeping beneath the stars. Nintendo trumps philately trumps homework. Homework is boring. Sorting stamps is boring (could it really have consumed the whole brain, as it seemed, all those years ago?). That book is boring, if you cannot sit still with it for fifteen minutes; yet your ancestors would have walked miles for the privilege of borrowing it from a library. Still, by recognizing these unconscious, minute-by-minute choices, you find that you can reorganize them, balance different styles of attention-grabbing, weigh the short term against the long term. Just as you manage to invest a painful half-hour on the StairMaster for the sake of muscle tone, you discover that you would rather fall asleep with the strains of a Mahler adagio echoing in your head than with those incessant colored Tetris blocks crashing through your dreams. Or can you multitask those, too?

Our idea of boredom—ennui, tedium, monotony, lassitude, mental doldrums—has been a modern invention. The word *boredom* barely existed even a century ago. To *bore* meant, at first, something another person could do to you, specifically by speaking, too long, too rudely, and too irrelevantly. Boredom as silence, as emptiness, as time unfilled—was such a mental state even possible? Samuel Johnson, in the eighteenth century, tried hard to believe it was not, for curious creatures such as ourselves. "To be born in ignorance with a capacity of knowledge," he wrote, "and to be placed in the midst of a world filled with variety, perpetually pressing upon the senses and irritating curiosity, is surely a sufficient security against"—here no simple word came to his mind—"the languishment of inattention."

The literary theorist Patricia Meyer Spacks, studying boredom through the centuries, retorts that Johnson protests too much. "Human beings need not languish," she comments. "And yet, perversely, they do. Minds feel vacant, hours seem long."

Maybe boredom is a backwash within another mental state, the one called *mania*—defined by psychologists as an abnormal state of excitement, encompassing exhilaration, elation, euphoria, a sense of the mind racing. Maybe our hurry sickness is as simple as that. We—those of us in the faster cities and faster societies and faster mass culture of the technocratic dawn of the third millennium C.E.—are manic. The symptoms of mania are all too familiar: volubility and fast speech; restlessness and decreased need for sleep; heightened motor activity and increased self-confidence. Of the possible mental illnesses, mania does not sound like the worst. Anyway, without mania, no boredom? These are the time obsessions of complex civilizations, populous nation-states with many technologies. In other forms of human society time passes differently. A few people still live as all our ancestors did, in small groups of hunter-gatherers, for example. Or, having domesticated plants and animals, people organize time around their duties to these. For example, John S. Mbiti sees the day of the Ankore of Uganda "reckoned in reference to events pertaining to cattle." His time-use chart makes the point:

6 A.M.: milking time.
12 noon: "time for cattle and people to take rest."
1 P.M.: draw water.
2 P.M.: cattle drink.
3 P.M.: cattle start grazing again. . . .

No 1,440 minutes here. And no mania. When the economist Juliet Schor was trying to bolster her case that modern industrialized peoples work longer hours than ever before, she argued that people did not work much in medieval or ancient times. Athenians had fifty or sixty holidays, she noted. True, and they also had slaves. "Primitives do little work," Schor wrote. "By contemporary standards, we'd have to judge them extremely lazy. If the Kapauku of Papua work one day, they do no labor on the next." Lazy! The Kapauku as couch potatoes? It is hard to imagine an appropriate methodology for counting the work hours of human populations on the edge of caloric subsistence, hunting or farming with bare hands and whatever tools could be fashioned of wood and stone. Their economies did not free enough time to support specialized occupations such as economist, writer, or time-use researcher. Before our enslavement by wristwatches and alarm clocks, the boundary between work and nonwork, between time on the clock and time off the clock, was fuzzier. Comparing work in Western societies of the 1990s to work in the same societies a mere twenty years earlier is difficult enough. Even now, an alien anthropologist watching someone sweat on an exercise bike and later swill Chardonnay at a business lunch might have trouble guessing which activity goes in the Toil column and which counts as Leisure.

Is there time for boredom to set in? Does boredom exist? Mbiti argues that time is just different in this culture. The most fundamental understanding of time, or a sense of time that precedes understanding, can only be hinted at with some of the words we commonly attach to time. He attaches some, to try to explain what time is for a culture like the Ankore's. In technological society, we *use, sell,* and *buy* time. In African life, a person *creates, produces,* and *makes* time—"as much time as he wants." It is possible in technological society to *waste* time. So Westerners, viewing an apparently idle African through the wrong lens, fail to see what time means here: "Those who are seen sitting down, are actually *not wasting* time, but either waiting for time or in the process of 'producing' time."

Waiting for time? Even better, producing time? What harried citizen of a technological culture could resist the seductive appeal of this prospect? All we have to do is think differently, and then, as we sit idle, watching the clouds, we might become little factories, manufacturing time for ourselves. All the time we need, all the time there is.

In that case we are not manufacturing a thing that can be traded for money; this time is not money. Nor are we are manufacturing any part of the space-time fabric in which we sit. We are just manufacturing life, as we live it.

The End

Your sense of acceleration has not blinded you to the brevity of the present moment. Oh, perhaps sometimes we allow ourselves to think we have a long history, but mostly we know that the peculiar hastening of our culture and ourselves has occurred in an instant. Any of the usual defined eras in human history—the Middle Ages, say, or the Ming dynasty—lasted many times longer than has the technocracy in which we live. And in those eras, by comparison, not much happened. We know that, however we define *modern* times, these times have been a mere eyeblink; that all recorded history has been just an eyeblink in the lifetime of the species; that *Homo sapiens,* information gatherer, has lived for barely an eyeblink in the history of terrestrial fauna . . .

So we are surrounded by ephemera. By any cosmic clock, every word that has been written, every song that has been recorded, every machine that has been devised, and even the paintings, buildings, and monuments we leave behind came into existence in the last instant. Everything is so new. We cannot help but think of it as short-lived. We don't have to be paleontologists to understand Stephen Jay Gould when he says, sternly: "If we continue to follow the acceleration of human technological time so that we end in a black hole of oblivion, the Earth and its bacteria will only smile at us as a passing evolutionary folly." Lately the accident of the Western calendar, nearing a round number of years in the common era, has heightened our impression of things as fleeting and transitory. Why else would so many scholars and authors be announcing "The End"? These are a few of the books that have helped us live out the last years of this millennium: *The End of Affluence. The End of Desire. The End of Economic Man. The End of Education. The End of Equality. The End of Expressionism. The End of Fame. The End of History. The End of Ideology.* We won't even consider the "Death of." Other titles have announced the end of Acting, Aging, the Alphabet, Architecture, Art, Beauty, Bureaucracy, Capitalism, Certainty, Christendom, the Church, Concerts, Days, Democracy, Innocence, Kinship, Knowing, Laissez-faire, Liberalism, Magic, Masculinity, Modernity, the Nation State, Nature, Parliamentary Socialism, Patriarchy, Physics, Print, Racism, Reform, Sanity, Science, Silence, Society, Sorrow, Vandalism, Welfare, and Work. A bit more vaguely, we have also had the end of the Rainbow, the Road (many), the Search, "It," the Story, the Line, and, of course, the World. None of these things has really ended, nor will they any time soon, so what is going on? What are these premature obituaries trying to tell us, and why are we so eager to hear the message?

Sometimes the "End of" announcements are just exaggerated ways of talking about fast change. Here comes the end of physics *as we know it,* we short-sighted folk; maybe particle physicists, whose careers have focused on expensive accelerators for several years now, are beginning to work on different problems entirely or finding it easier to get jobs on Wall Street. The end of history might, on close scrutiny, turn out to be nothing more than the end of the Cold War, a transitory state of international affairs that lasted a few decades. The end of work might turn out to be a decline in the proportional share of manufacturing employment in advanced Western economies. Still, all these titles suggest a kind of destiny in human affairs, a one-way path toward fulfillment or climax, and not a timeless, cyclical, ever-changing form of history.

You have seen the following graph. You have seen it more than once.

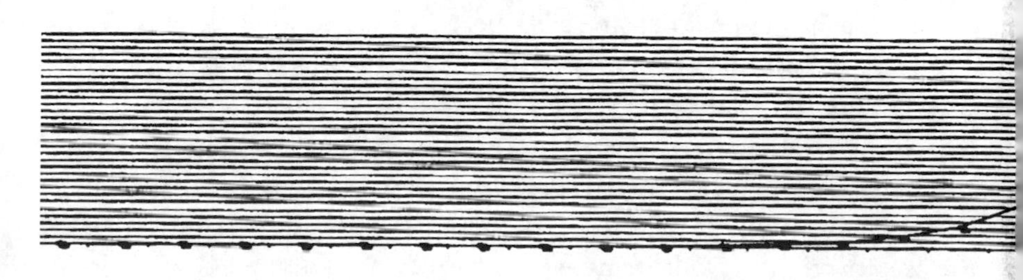

It depicts the long-threatened population explosion, or some kind of population explosion, plotted over a few centuries, or millennia, or any time scale at all. It represents the growth in computer ownership over the last two decades. The number of commercial Internet hosts rising over a mere four years. Software patents granted from 1971 to the present. Chest-pain emergency departments in the 1990s. Millions of instructions per second carried out by a matchbook-size computer. Potential sexual partners. Mustards. Published words. Four-minute milers. Everything, it seems, that grows out of the interaction between human beings. The amount of stuff to do, divided by the amount of time available.

This proliferation of choice represents yet another positive feedback loop—a whole menagerie of such loops. The more information glut bears down on you, the more Internet "portals" and search engines and infobots arise to help by pouring information your way. The more telephone lines you have, the more you need. The more patents, the more patent lawyers and patent search services. The more cookbooks you buy or browse, the more you feel the need to serve your guests something *new;* the more cookbooks you need. The complications beget choice; the choices inspire technology; the technologies create complication. Without the distribution and manufacturing efficiencies of the modern age, without toll-free numbers and express delivery and bar codes and scanners and, above all, computers, the choices would not be multiplying like this. If a graph can be a cliché, the graph for exponential growth has become a cliché.

Not only does all exponential growth look the same, once the scale is properly adjusted, but it looks scary. For all that time, something is at zero, or one, or some number small enough to stay off the screen. Then it starts growing. All too soon the growth becomes impossibly fast. At the end, it rises like a wall—the limit, the present, right now. Is it any wonder we develop a sense of our future as a thing curtailed? We look at these graphs, we note our present location just before the end, and we shudder at the impossibility of continuing to scale those heights. Surely we will hit limits imposed by physical laws. The computer hard drive—in 1956 the size of a small truck, now storing billions of bytes on ever-tinier, ever-faster disks—can't shrink forever, can it? Can the economy's capacity to absorb more specialized broadcast-television channels and more whimsically named bottles of hot sauce grow forever? Maybe when the slope ahead gets impossibly steep, acceleration gives way to paralysis. Certainly there is no biological reason for us to accept it comfortably. "Humans endure a more or less confined life, far removed from the hurried pace of exponentials," notes Greg Blonder. "Forty-five Fahrenheit is cold, eighty-five Fahrenheit is warm. Five hundred calories a day, you starve; three thousand a day, you're as fat as a pig." So much for our unjetlagged origins. By contrast, "Exponentials start slowly and remain disarmingly out of sight. Yet they build strength relentlessly until they've grown too large to ignore. By then, whole industries have changed and whole cultures have fallen." Of course, our culture is, self-consciously, the one through the looking glass. Alice thought that running very fast for a long time would get you to somewhere else. "'A very slow sort of country!' said the Queen. 'Now, *here,* you see, it takes all the running you can do, to keep in the same place.' "

We make choices. But we have a sense that our choosing is not entirely free. We're like unvaccinated travelers through territory awash in disease. At any moment we can catch that fever—and the fever feels good, at first. We can pull out that cellular phone at the beach. We can reach for the remote control or head for the drive-through.

The voices reminding us of the dangers of speed are many. Some are inside our heads. Walk, don't run. Relax. Simplify. Let the phone ring. Beware Type A. "I've always moved at a fast clip," confesses Jay Walljasper dramatically, introducing a special "Slow Down" edition of the *Utne Reader.* "I can't stand small talk, waiting in line, or slow numbers on the dance floor. . . . It has gotten to the point where my days, crammed with all sorts of activities, feel like an Olympic endurance event: the everydayathon. . . . I hear an invisible stopwatch ticking even when I'm supposed to be having fun." Sure. We catch the fever. We choose mania over boredom every time. "The historical record shows that humans have never, ever opted for slower," points out the historian Stephen Kern. We fool ourselves with false nostalgia—a nostalgia for what never was. Whenever we speed up the present, as a curious side effect we slow down the past. "If a man travels to work on a horse for twenty years," Kern says, "and then an automobile is invented and he travels in it, the effect is both an acceleration and a slowing. . . . That very acceleration transforms his former means of traveling into something it had never been—slow—whereas before it had been the fastest way to go." Until the futurist Filippo Marinetti began talking about speeding up rivers, "the Danube had never seemed so deliciously slow." Peering back through history, we see scenes in a kind of slow motion that did not exist then. We have invented it.

Civilization has created one kind of quarantine ward, with walls the fever of speed does not penetrate. This is the place where people *do* time; where their most fervent desire is to *pass* time and *kill* time. In prison there are short-timers and long-timers, people doing hard time and people doing harder time. The prisoner's curse is boredom, and yet not exactly boredom. It is a powerlessness to control time; the control of time, a precious possession, has been taken away. "Inside the prison walls history comes to a halt; time's mechanism goes awry," writes Maurice Lever, biographer of the Marquis de Sade. "The prisoner is suddenly plunged into 'uchronia,' into a world where time does not exist." Sade, like so many famous prisoners—from Socrates to Hitler—found new resources and focus upon turning inward. Sade wrote, sometimes in blood. Malcolm X copied a dictionary and read. "Months passed without my even thinking about being imprisoned," he said. "In fact, up to then, I never had been so truly free in my life." Prisoners regularly say that short sentences are harder than long ones. The difference between waiting and living, perhaps. Alexander Solzhenitsyn, returning to freedom, found himself more bored waiting sixteen minutes for a trolley bus than he had been in the gulag during empty sixteen-hour days, devoid of any event worth recalling. Outside the prison walls, sixteen minutes can indeed seem long. In the same century, on the same planet, Tom Parsons, a retired teacher working toward a doctorate in Auckland, New Zealand, found his own psyche transformed—spoiled for mere "monopursuits"—so absorbed was he in, as he said, "multitasking, or at least rapid task-switching, from: following Wall Street, to reading/writing E-mail, to participating in online conferences, to reading world news from several sources, to writing down fragments of research ideas, to housekeeping on the disk drive, to glancing at the latest satellite weather pic, to improving my programming skills." It turns out that multitasking has been our destiny all along—not killing time, not doing time, but mastering time.

We live as free men and women, so we show up on time, we mark time, we worry about time, we time ourselves, partly for the simple reason that we can. We multitask because we can. If the minute hands, or even the second hands, could be legislated off our watches, we would suffer. We might relax, but we would suffer anyway.

We do feel the rush of time more as we grow older. Then, time does go by faster for us. Perhaps that is partly because the end is nearer. Psychologists have isolated a "gradient of tension" to measure the shift in our sense of time as we approach a critical point—the end of a baseball game, a journey, a book, a millennium, a lifetime. Behind all our haste, all that migraine-like pressure to hurry, lurks the fear of mortality. But perhaps the sense of speed comes also from having experienced more. People accumulate responsibilities and time fillers as ocean piers accumulate barnacles.

You are aware that the director of the Directorate of Time is something of a philosopher. He has written, "We experience time intervals as much shorter than when we were young." He even has equations for this: "Delta t(s) ~ Delta Exp/Total Exp" and "dt(s) ~ dt/t or integrated t(s) ~ ln(t)," by which he means, the more we have experienced, the faster time flows. Depressants like alcohol slow time, because the brain receives fewer inputs per second. You may feel, as so many do, that your life could be plotted on a scale where the years from age ten to age twenty seem as long (as event-full) as the years from age twenty to age forty or from forty to eighty. Exponential growth at its most damning. On this scale, the moment of birth is at negative infinity, and as for death . . . someone else might quote Woody Allen, but the director favors Epicurus: "Death is nothing to us, since when we are, death has not come, and when death has come, we are not."

Death may be an absolute but time is not. Our ancestors may have considered time to be divine property, but we know better—we who have created jet lag, slow-motion instant replays, methamphetamines, the International Date Line, the relativity of physicists, leap years and leap seconds. Come to think of it, Winkler is not really setting the pace—not for you. Synchronize your watch according to his clocks, sure, but you will serve as your own director of your own time directorate. Even if you feel yourself rushed by the sheer plenitude of things, even if you eat when the clock says to, you can remember that time is defined, analyzed, measured, and even constructed by humans. It may help to think of time as a continuous flow, rather than a series of segmented packages. Or to find aggressive ways of squandering the time you save. Or at least to recognize that neither technology nor efficiency can acquire more time for you, because time is not a thing you have lost. It is not a thing you ever had. It is what you live in. You can drift in its currents, or you can swim.

The director has finished, it seems. But you cannot resist asking a few questions about the psychological motivation of a timekeeper with such profound responsibilities. He cooperates: "Accuracy, precision, control—this is something which is to me aesthetically pleasing."

Are you a punctual person?

"I try to be."

What kind of watch do you wear?

"None."

Why is that?

"I don't need to. This would be an admission of defeat."

Defeat! Whatever can he mean? Anyway, a reasonably accurate clock hangs on the wall just behind your left shoulder, and you see Winkler glance at it. Your half-hour is up.

Afterword to the Vintage Edition

The book—even a book called *Faster*—has become a notoriously slow device: slow in the writing, slow in production, slow to read and absorb. That was not always true. The printed word began as advanced technology for rapid transmission of data into the brain. In terms of bits per second, there was no better way to get information, or a story, or facts, from *out there* to *in here.*

Nowadays, by most modern data-processing standards, the bits on the printed page stream by slowly. It takes a long time to set them down in the first place. Readers may, if they wish, treat books as a sort of random-access memory, to be dipped into here and there, but that's not how they're designed (cookbooks and encyclopedias aside). Books convey data sequentially, if not linearly; they mean to deliver their news one piece after another, gradually building up a sort of structure across time. Books may either ignore or use to advantage the subtle and fickle nature of the reader's memory. Anyway, reading a book is meant to be slow and focused, in contrast to the more modern style of reading we are learning from our peregrinations through cyberspace.

Meanwhile, for reasons shrouded in the history of institutions, it takes a very long time to publish most books. Thus, by the time *Faster* started appearing on bookstore shelves, it was already out of date. Certain details had a quaint ring. It was, after all, the work of a past millennium.

So the chapter on "Real Time" considers the phenomenon called day trading. When I wrote this section, day trading was a new form of madness. I had to define "day traders"—people buying and selling stock on a time scale measured in minutes, rarely, if ever, holding a position overnight. I had to offer evidence of their existence—from students online in their dorm rooms to retirees from actual, productive jobs. I estimated their number as hundreds or thousands. I elicited from the chairman of the Securities and Exchange Commission, Arthur Levitt, only a mild comment to the effect that short-term thinking did not serve investors well.

Then, in an eyeblink, day trading went from an obscure sideshow to a powerful force in the world's financial markets. The markets have never been so volatile, and the volatility, by no coincidence, finds its greatest extremes in the technology and Internet stocks most closely monitored by adrenaline-charged day traders at their computer screens. It is starting to seem as though the markets have undergone a final divorce of the prices of shares from any long-term sense of company value. "Does this craziness do any harm?" asks the economist Paul Krugman after a particularly severe market drop. "Or is it all audio and video, signifying nothing?" No one knew, and in the very next week, the five thousand stocks comprising the Nasdaq Composite Index lost a quarter of their value—for no reason that any analyst or regulator could perceive. It used to be rare for the Nasdaq to rise or fall by a full percentage point in a day; in 1997 that happened one day in three, an all-time record. Now it happens almost every day. As of mid-2000, moves of one percent or more occur three days in four. One day the Nasdaq lost 13 percent in just three hours.

A word often applied to financial markets is *efficient.* An efficient market reflects all available information and expectations, and does this quickly. For a stock's price to be seriously out of line with reality, there must be some inefficiency, some information unequally shared. An efficient market tends toward stability. Or so we once thought. Now we know better. Nothing creates volatility more wondrously than the efficient flow of information—news traveling faster and faster, to everyone at once. When everyone knows everything, and everyone agrees, then everyone will try to buy or sell together—instantaneously, if they can (and in a day-trading world, they *can*). Volatility will become infinite. Aficionados of chaos will know why. Efficiency does *not* imply equilibrium. What little stability remains in the markets reflects our last vestiges of *in*efficiency—our gaps in knowledge, our ornery human differences in judgment, and our few moments offline. "When things are going well there is a strong tendency to suppose that the financial markets can take care of themselves," Krugman commented. "Well, they can't."

Nor can Rush Limbaugh, it seems. Here and there *Faster* touches on the spread of digital speech compression, a technology that accelerates the spoken word by winnowing the pauses and *um*'s and *ah*'s. Here and there, too, the book delves into the not-so-obvious role that pauses play in the various threads of our lives—pauses on scales of days, hours, or even fractions of a second. The right-wing radio star was shocked nonetheless to discover broadcasters nationwide using speech compression to speed up his daily live program, ever so slightly, without making him sound like the Chipmunks, and using the time gained to sell extra commercials. "In life everything moves faster now," Kraig Kitchin, the radio executive distributing Limbaugh's show, told a *New York Times* reporter in January 2000. "You get more e-mails and voice mails than you can keep up with. But you still have the same number of hours in the day." Why, yes!

The technology is insinuating itself into the world's broadcast streams more rapidly than I imagined possible. On Limbaugh's show, a box retailing for $12,000 and filtering the signal in real time finds so much "redundancy" to chop from Limbaugh's diction that stations manage to save up to six minutes an hour. The device "creates additional time," the manufacturer claims boldly. It stores the flowing bits of sound in a buffer; analyzes them on the wing, and chops as much as a tenth from each passing second. Then it releases the slimmed-down signal.

Limbaugh got over his shock quickly enough; he shares in the advertising revenue, after all, and listeners really couldn't tell. "Of course, to the technology nerds, this is a fascinating device," he said. Speech compression is becoming ubiquitous. Few news or interview shows on radio or television broadcast people's words in gross, unfiltered masses anymore, with all our natural halting and stammering. The spoken word rattles right along, and the broadcasters sell the time they save. How sluggish, by contrast, our own paltry voices come to seem. The sounds and sights of the real world compete with these digital, virtual realities.

So we are trapped yet again by desire for efficiency and need for equilibrium; our recollections of silence and our love of cacophony. Cyberspace, especially, draws us into the instant. Messaging is instant. Overnight delivery is slow. We measure in minutes and seconds the wait for headline news, credit card approval, romance, and wisdom. Some readers expected from *Faster* a fast read; others a concise answer to a set of questions we have barely learned to ask. I fear they may be disappointed. It has its Web site, but it's still meant to be a book—and slow. There's no revolution without counterrevolution (that's one of Newton's laws, isn't it?), and every day new signs of resistance—if not outright panic—arise over our accelerating world. In *Faster* I touched on such phenomena as the Slow Food Movement and the Simplify-Your-Life movement (and the Simplify-Your-Life information glut). One could just as well point to the Long Now Foundation and the Society for the Deceleration of Time. Time management and time coaching have become giant enterprises. Many early readers of *Faster* took pains to let me know that they do not wear wristwatches. Strange—sales figures from the wristwatch companies don't yet show any signs of decline.

I believed when I began *Faster,* and believe now more than ever, that we are reckless in closing our eyes to the acceleration of our world. We think we know this stuff, and we fail to see connections. We're perennially surprised—by our friends, by our government, by ourselves. We struggle to perceive the process of change even as we ourselves are changing. After all, flux is our style, if not our destiny. We don't exist in a steady state, and we don't have a motionless platform from which to observe the changing world around us. Sometimes we fail to perceive profound transformations that we've been staring at; sometimes we blink and we notice a revolution. The most profound comment on this is still Richard Feynman's; he was sitting outdoors in New Mexico, looking up at a blue and turbulent sky and talking about the evolution of his field, theoretical physics. "It is really like the shape of clouds," he said. "As one watches them they don't seem to change, but if you look back a minute later, it is all very different."

Blank page 288

Acknowledgments and Notes

The following notes provide citations to published sources. When a place is described or a person is quoted and I *don't* supply a source note, it means I was there (Directorate of Time, movie-production set, airline control center, telephone headquarters, etc.), or someone spoke or wrote to me directly, or I heard those words and wrote them down in (pardon the expression) real time. I owe a debt of thanks to the many people who shared their time with me during the reporting of this book, helping me with interviews and archival research. I'm also in debt to the hundreds of people who responded on-line to my requests for assistance; only a few of them are quoted or named in the text, but all were helpful.

A bibliography on time and speed could fill a volume in itself. Apart from those cited in the notes, I will just list a few excellent books with strikingly varied approaches to some of the issues I try to explore:

J. T. Fraser, ed., *The Voices of Time* (Amherst: University of Massachusetts Press, 1981).

Sebastian de Grazia, *Of Time, Work and Leisure* (New York: Vintage, 1993).

Stephen Kern, *The Culture of Time and Space: 1880–1918* (Cambridge: Harvard University Press, 1983).

Robert Levine, *A Geography of Time: The Temporal Misadventures of a Social Psychologist* (New York: Basic Books, 1997).

John P. Robinson and Geoffrey Godbey, *Time for Life: The Surprising Ways Americans Use Their Time* (University Park: Pennsylvania State University Press, 1997).

G. J. Whitrow, *Time in History* (Oxford: Oxford University Press, 1988).

For their direct assistance, insight, and wisdom, I thank Jack Rosenthal of *The New York Times Magazine,* my agent Michael Carlisle, Greg Blonder, Jerome Chou, David Feldman, Cathy Fuerst, Beth, Donen, Betsy, and Peter Gleick, Douglas Hofstadter, Uday Ivatury, Zachary Katz, Judith Kogan, Nicholas Lemann, Frida Lindman, Martin Seligman, William Slaughter, Betsy Stark, Joseph Straus, Edward Tenner, Laura Tolkow, Roben Torosyan, Craig Townsend, Peg Tyre, Hugh Wolff, David Wyner, and Martha Zornow. My editor for this book, as for its predecessors, was the incomparable Dan Frank.

The whole thing was Cynthia Crossen's idea. Beyond that, even to begin thanking or acknowledging her requires words that I don't have.

I'M GOING TO KILL MYSELF Woody Allen, "Everyone Says I Love You," screenplay, 1996.

CLOCKS CANNOT TELL W. H. Auden, "No Time," in *Collected Shorter Poems, 1927–1957* (New York: Vintage, 1975).

Pacemaker

LONG AGO A FOUR-FOOT BALL Ian R. Bartky and Steven J. Dick, "The First Time Balls," *Journal for the History of Astronomy* 12 (1981): 155–64; Ian R. Bartky and Steven J. Dick, "The First North American Time Ball," *Journal for the History of Astronomy* 13 (1982): 50–54; Gernot M. R. Winkler, "Changes at USNO in Global Timekeeping," *Proceedings of the IEEE* 74 (January 1986): 151.

SPEED IS THE FORM OF ECSTASY Milan Kundera, *Slowness* (New York: HarperCollins, 1996), pp. 2 and 1.

HURRY SICKNESS Among others, Barbara Allen, *Timewatch: The Social Analysis of Time* (Cambridge, Mass.: Polity Press, 1995), pp. 53 ff; Larry Dossey, *Space, Time, and Medicine* (London: Shambala, 1982), p. 50.

THE WATERFLEA Theodore Zeldin, *An Intimate History of Humanity* (New York: HarperCollins, 1995), pp. 352 f.

TECHNOLOGY HAS BEEN A RAPID Ibid.

NEW YORKER CARTOONS Donald Reilly, *New Yorker,* November 25, 1996, p. 89; Richard Cline, *New Yorker,* January 27, 1997, p. 77.

IT SEEMS LIKE THE WHOLE "Diary, by Bill Gates, CEO of Microsoft" (http://www.slate.com/Code/DDD/DDD.asp?file=billg&iMsg=2, March 12, 1998).

I'LL TRY TO BE BRIEF *The Late Show with David Letterman,* February 12, 1999.

OPENING LOCKERS, GRABBING April Tomaino, "Four Minutes Not Enough Time to Go to Locker, Restroom, Etc." (http://168.216.219.18/bab/jan95/fourmins.htm).

YOUR LIFE IS LIVED Mark Helprin, "The Acceleration of Tranquillity," *Forbes,* December 2, 1996, pp. 15 ff.

Life as Type A

TYPE A PERSONALITIES HAVE Douglas Coupland, *Microserfs* (New York: HarperCollins, 1995), pp. 276–77.

A HARRYING SENSE OF TIME Meyer Friedman and Ray H. Rosenman, *Type A Behavior and Your Heart* (New York: Knopf, 1974), p. 4.

I DROVE ALL THE WAY Marilyn M. Machlowitz, "A New Take on Type A," *New York Times,* March 3, 1987, sec. 6, p. 40.

DIRECT YOUR ATTENTION Laura Mansnerus, "Count to 10 and Pet the Dog," *New York Times,* April 25, 1993, sec. 6, p. 74.

THE GREAT MEN Cecil Webb-Johnson, *Nerve Troubles: Causes and Cures* (London: Methuen, 1929), p. 5.

THE STUDY THAT STARTED Meyer Friedman and Ray H. Rosenman, "Association of Specific Overt Behavior Pattern with Blood and Cardiovascular Findings," *Journal of the American Medical Association* 169 (March 21, 1959): 1286–96.

LOWER BLOOD PRESSURE E.g., David J. Lee, Orlando Gomez-Marin, and Ronald J. Prineas, "Type A Behavior Pattern and Change in Blood Pressure from Childhood to Adolescence. The Minneapolis Children's Blood Pressure Study" (University of Miami School of Medicine), *American Journal of Epidemiology* 143 (1996): 63.

HYPERVIGILANCE V. A. Price, *Type A Behavior Pattern: A Model for Research and Practice* (New York: Academic Press, 1982).

FEWER DAYDREAMS Cynthia Perry, "Sustained Attention and the Type A Behavior Pattern: The Effect of Daydreaming on Performance," *Journal of General Psychology* 119 (July 1, 1992): 217.

PETLESSNESS "The Health Benefits of Pets," workshop summary, September 10–11, 1987, National Institutes of Health, Office of Medical Applications of Research, Bethesda, Md.

A FEW DOZEN PRESCHOOL "Type A Preschoolers," Science Watch, *New York Times,* June 7, 1988, p. C7.

FREE-FLOATING, BUT WELL Friedman and Rosenman, *Type A Behavior and Your Heart,* p. 88.

MUNICIPAL CLERKS' AND Ibid., p. 61.

ALREADY 6:20, AND BOOKS Cullen Murphy, "Type B, A-Wise," in *Just Curious* (Boston: Houghton Mifflin, 1995), p. 195.

The Door Close Button

ARCHETYPAL DUMB EMBEDDED The on-line hacker Jargon File, version 3.1.0, October 15, 1994. Many anonymous authors.

WE'LL CALL ELEVATOR MONITORING George R. Strakosch, "A New Dimension in Elevator Monitoring," *Elevator World* 42, no. 8 (1994): 126–29.

BROKEN EARDRUM James W. Fortune, "Mega High-Rise Elevators," *Elevator World* 43, no. 7 (1995): p. 63.

DISTORTED TIME SENSE George R. Strakosch, *Vertical Transportation: Elevators and Escalators,* 2d ed. (New York: John Wiley, 1983), p. 444.

WAITING, SOME STAND STILL Spivack Associates, *The Sourcebook: Human Behavior and Perception in Elevators,* report for the Otis Elevator Company (Cambridge, Mass.: Spivack Associates, 1979), pp. 115–20.

AIR-BORNE PREDATORS Ibid., pp. 120–21.

NEW YORK CITY'S SUBWAY Richard Pérez-Peña, "Transit Authority Urges Platform Etiquette to Speed Subways," *New York Times,* November 12, 1996, p. B1. James Rutenberg, "TA Aims Not to Say Please/Part of Speedup Drive," *New York Daily News,* March 29, 1999.

THEY KNOW THE ATTENDANT Ralph A. Weller, "Autotronic 'Without Attendant' Elevator Presentation Using the Otis Electronic Demonstrator Model," September 1, 1953, p. 6. Otis Elevator Company Historical Archives.

Your Other Face

YOUR WATCH PROCLAIMS Radio advertisement for Tourneau, 1998.

BUTTONS ARE CHEAPER AND Donald A. Norman, *The Design of Everyday Things* (New York: Doubleday, 1988), p. 31.

THE CLOCK IS THE FIRST *Karl Marx and Frederick Engels, Selected Correspondence 1846–1895,* trans. D. Torr (New York: International Publishers, 1942), p. 142.

IN THE MECHANIZED FACTORY Sebastian de Grazia, *Of Time, Work and Leisure* (New York: Vintage, 1993), p. 303.

THE CHINESE TREATED TIME David S. Landes, *The Wealth and Poverty of Nations: Why Some Are So Rich and Some So Poor* (New York: Norton, 1998), p. 50.

IT'S BEEN 13 DAYS Avodah Offit, *Virtual Love* (New York: Simon & Schuster, 1994), p. 31.

NEVER LOSE ANOTHER Peter Monta, "How to Set Your Wristwatch to Extremely High Accuracy" (http://image.mit.edu/~monta/hos/wristwatch.html).

THE IDIOTIC FASHION Helmut Kahlert, Richard Mühe, and Gisbert L. Brunner, *Wristwatches: History of a Century's Development,* trans. Edward Force (West Chester, Pa.: Schiffer Publishing, 1986), p. 12.

PINE CONES IN MY STUDY J. T. Fraser, *Of Time, Passion, and Knowledge,* 2d ed. (Princeton: Princeton University Press, 1990), p. 67.

THE *RÉGLEURS,* THOSE ATHLETES David S. Landes, *Revolution in Time* (Cambridge: Harvard University Press, 1983), p. 346.

OUR QUEST FOR THE PRECISE Anthony F. Aveni, *Empires of Time* (New York: Basic Books, 1989), p. 100.

CLOCKMAKERS WHO MEASURED Landes, *Revolution in Time,* p. 133.

HOW VARIABLE SUBJECTIVE TIME Donald Lowe, *History of Bourgeois Perception* (Chicago: University of Chicago Press, 1982), p. 35.

ALTHOUGH THE TICKS Fraser, *Of Time, Passion, and Knowledge,* p. 67.

Time Goes Standard

THE CHOPPING UP OF TIME Quoted in Michael O'Malley, *Keeping Watch: A History of American Time* (New York: Viking, 1990), p. 136.

THE GODS CONFOUND THE MAN Titus Maccius Plautus, quoted in Aulus Gellius, *Noctes Atticae* 3.3.

HAVE NOT MEN IMPROVED Brooks Atkinson, ed., *Walden and Other Writings of Henry David Thoreau* (New York: Modern Library, 1992), p. 111.

YOU HAVE PRACTICALLY A David Elliot to Theo. Vail, January 10, 1920. AT&T Archives.

TIME REQUESTS ACCOUNTED FOR K. W. Watson to Mr. B. Gherardi, "Giving Out Information as to Time of Day" (memorandum), April 23, 1921. AT&T Archives.

THE WHOLE ENTERPRISE Edward Hungerford, "Time by Telephone," letter to the editor, *New York Times,* September 1, 1926.

NEW YORKERS PONIED UP *Headquarters Bulletin,* September 7, 1928, p. 3. AT&T Archives.

The New Accelerators

CAFFEINE, WE NOW KNOW, *Desk Reference to the Diagnostic Criteria from DSM-3-R* (American Psychiatric Association, 1987).

COFFEE À GO GO http://www.illuminatus.com/fun/agogo/caffeine.html

WHAT COMES TO MIND Wolfgang Schivelbusch, *Tastes of Paradise: A Social History of Spices, Stimulants, and Intoxicants,* trans. David Jacobson (New York: Pantheon, 1992), pp. 111, 115.

WE FLEW ON THE WINGS Quoted in James A. Ward, *Railroads and the Character of America, 1820*–1887 (Knoxville: University of Tennessee Press, 1986), p. 110.

THE NEW ACCELERATOR In *Best Science Fiction Stories of H. G. Wells* (New York: Dover, 1966), pp. 165–76.

THE TIME MACHINE H. G. Wells, *The Time Machine: An Invention* (London: Heinemann, 1895).

FLATLAND Edwin Abbott Abbott, *Flatland: A Romance of Many Dimensions by a Square* (New York: New American Library, 1984).

TIME'S CATTLE DRIVE Nicholson Baker, *The Fermata* (New York: Vintage, 1995), p. 47.

STOPPING THE UNIVERSE Ibid., p. 190.

I'VE LOST ALL CONCEPTION Ibid., p. 295.

Seeing in Slow Motion

IT KICKS THE THEORY OF VISION *Best Science Fiction Stories of H. G. Wells,* p. 169.

WE PREDICT THAT HIS INSTANTANEOUS *Alta California,* May 5, 1880.

WE DOUBT WHETHER THE CONTRIBUTION *The London Globe,* quoted in Gordon Hendricks, *Eadweard Muybridge: The Father of the Motion Picture* (New York: Grossman, 1975), p. 206.

ON ACCOUNT OF THE PERSISTENCY Quoted in L. Moholy-Nagy, *Vision in Motion* (Chicago: Paul Theobald, 1947), p. 247.

THE GESTURE WHICH WE WOULD Umberto Boccioni, Carlo Carrà, Luigi Russolo, Giacomo Balla, and Gino Severini, "Technical Manifesto of Futurist Painting," *Poesia* (Milan), April 11, 1910.

HAROLD EDGERTON Harold E. Edgerton and James R. Killian, Jr., *Moments of Vision: The Stroboscopic Revolution in Photography* (Cambridge: MIT Press, 1985); Estelle Jussim and Gus Kayafas, *Stopping Time* (New York: Abrams, 1987); Stephen Dalton, *Split Second* (Salem, N.H.: Salem House, 1984).

FEMTOSECOND FLASHLIGHT Ian Walmsby and Rick Trebino, "Measuring Fast Pulses with Slow Detectors," *Optics & Photonics News* 7, no. 3 (March 1996): 23.

WE DECLARE THAT THE SPLENDOR Filippo Tommaso Marinetti, *Futurist Manifesto,* 1909.

NATURE REFUSES TO REST John Updike, *Toward the End of Time* (New York: Knopf, 1997), p. 4.

In Real Time

WE'RE FIGURING IT OUT Farzad Dibachi, quoted in Lawrence M. Fisher, "Sun Microsystems to Buy Diba, a Start-Up," *New York Times,* August 1, 1997, p. D4.

THIS IS A TOUGH PRACTICE Louis Menand, "But Is It Good for the News?" *New Yorker,* November 18, 1996, p. 8.

WE PREDICT THAT LARGE-SCALE Bernardo A. Huberman and Tad Hogg, "Phase Transitions in Artificial Intelligence Systems," *Artificial Intelligence* 33, no. 2 (1987): 155–71.

THE WHOLE THING HAS BEEN CAUGHT Jeff Greenfield on CNN, November 19, 1998.

CNN'S GOING TO BE ON Peter Jennings, interviewed November 18, 1998, by Terry Gross, "Fresh Air," National Public Radio.

THE CASINO CAPITALISTS "Why Internet Shares Will Fall," *The Economist,* January 30, 1999.

SPEED IS GOD, AND TIME John Markoff, "A Quicker Pace Means No Peace in the Valley," *New York Times,* June 3, 1996, p. D1.

FIVE-YEAR PLANS??? Quoted in Brad Wieners, "Managing on (Internet) Time," *Wired,* June 1998, p. 86.

BY 1997, TWENTY-FOUR MONTHS Noelle Knox, "G.M. Speeds Up Car Development," *New York Times,* March 20, 1998, p. F1.

A STUDY WIDELY QUOTED Christopher Meyer, *Fast Cycle Time* (New York: Free Press, 1993), p. 48.

Lost in Time

THE RATE OF CHANGE WILL BE James Burke, *The Pinball Effect* (Boston: Little, Brown, 1996), p. 5.

RETREAT FROM INTENSITY Edward Tenner, *Why Things Bite Back: Technology and the Revenge of Unintended Consequences* (New York: Knopf, 1996), p. 276.

On Internet Time

LAWYERS ARE TEDIOUSLY SLOW *Modern Equipment Makes the Lawyer Money,* American Bar Association Economics of Law Practice Series Pamphlet No. 3 (Minneapolis: West Publishing, 1958).

BEEPER MEDICINE E. Ide Smith and William P. Tunnell, "Beeper Medicine," *Southern Medical Journal,* July 1988, p. 816.

JAKOB NIELSEN http://www.useit.com/alertbox/9710a.html; http://www.useit.com/alertbox/whyscanning.html; and personal communication.

PRETEND YOU HAVE JUST "37 Smart (Not Silly) Ways to Simplify Your Life," *Redbook,* November 1996.

WHEN ALL THIS IS DONE Customer comment, Amazon.com, July 21, 1997 (http://www.amazon.com/exec/obidos/ASIN/0786880007/o/qid=921183568/sr=2-1/002-6771736-6597229).

IT IS THE WAY OF KEEPING Herbert Stein, "The Sidewalk Nursery: The True Comforts of a Cell Phone or Walkman May Lie Deeper Than We Think," *Slate,* September 17, 1998 (http://www.slate.com/itseemstome/98-09-17/itseemstome.asp).

Quick—Your Opinion?

RIGHT ON TO THE AIR Cynthia Crossen, *Tainted Truth: The Manipulation of Fact in America* (New York: Simon & Schuster, 1994), p. 124.

THESE BAROMETERS MEASURE SOMETHING Ibid., p. 126.

IMAGINE DOSTOYEVSKY Quoted in Peter Applebome, "Round and Round in the Search for Meaning," *New York Times,* March 29, 1998, sec. 4, p. 1.

SOUND BITES HAVE GROWN SHORT Mitchell Stephens, *the rise of the image the fall of the word* (New York: Oxford University Press, 1998), pp. 145–46.

WHEN JOHN F. KENNEDY Tony Schwartz, *The Responsive Chord* (Garden City, N.Y.: Anchor Books/Doubleday, 1973), p. 42. Also, Everett M. Rogers, *Diffusion of Innovations,* 4th ed. (New York: Free Press, 1995), p. 75; Allan R. Pred, *Urban Growth and the Circulation of Information: The United States System of Cities, 1790–1840* (Cambridge: Harvard University Press, 1973).

OUR OWN INTELLIGENCE IS TIED Douglas R. Hofstadter, *Gödel, Escher, Bach: An Eternal Golden Braid* (New York: Basic Books, 1979), p. 679.

Decomposition Takes Time

A NUMBER OF MAGAZINE ADS Joe Keyser, "Mythbusting: Ten Tall Tales about Composting," November 18, 1998 (http://bbg.org/gardening/natural/compost/ten.html).

BEEN THERE LEATHER JACKET The Territory Ahead, mail-order catalogue, 1997.

ALL THE WHILE, WE'VE FELT THE SKY Arthur T. Winfree, *The Timing of Biological Clocks* (New York: Scientific American, 1986), p. 5.

THAT DISCONCERTING SENSATION Ibid.

PUT ON THE LIGHT-PULSE GOGGLES *Consumer Reports,* "Selling It," July 1993, p. 483.

IN OUR DAY OF ELECTRIC WIRES Mark Twain, *More Tramps Abroad* (London: Chatto & Windus, 1897).

On Your Mark, Get Set, Think

WE OFTEN HEAR THE CLICHÉ David Feldman, *Do Penguins Have Knees?* (New York: HarperCollins, 1991), p. 210.

MOST PEOPLE ARE ONLY ABOUT David G. Wittels, "You're Not as Smart as You Could Be," *Saturday Evening Post,* April 17, 1948, p. 20.

AT LEAST IN INDUSTRIES LIKE Nicholas Lemann, "A Fool's Goal," *Forbes ASAP,* November 30, 1998.

SOMETHING IN THE NATURE OF Charles Spearman, *The Abilities of Man* (New York: Macmillan, 1927), p. 5. Quoted in Stephen Jay Gould, *The Mismeasure of Man* (New York: Norton, 1981), p. 266.

UNIVERSITY STUDENTS SHOW Arthur R. Jensen, *Behavioral & Brain Sciences* 7 (1984).

THE SIMPLEST INTERPRETATION Edward M. Miller, "Intelligence and Brain Myelination: A Hypothesis," *Personality and Individual Differences* 17, no. 6 (December 1994): 803–33.

IF ANYTHING, THE ESSENCE In Sohan Modgil and Celia Modgil, *Arthur Jensen: Consensus and Controversy* (Brighton: Taylor & Francis, 1987).

A Millisecond Here, A Millisecond There

I AM TRYING T TO GET THE HANG Mark Twain, letter to Orion Clemens, December 9, 1874.

CARRIED THE GUTENBERG Marshall McLuhan, *Understanding Media: The Extensions of Man* (New York: McGraw-Hill, 1965), p. 262.

THE IDEAL PRINTER Winn L. Rosch, *Winn L. Rosch Hardware Bible* (Indianapolis: SAMS Publishing, 1997).

THE TELEPHONE SAVES TIME *Time-Saving,* monograph by the Bell Telephone Companies, Louisiana Purchase Exposition, St. Louis, 1904, p. K6. AT&T Archives.

1,440 Minutes a Day

The time-use statistics in this and the next few chapters require a special note. They come from records maintained by official or semi-official sources, including government agencies, research groups, telephone companies, and the like; from a wide variety of published studies; and from books and articles, principally including the following: John P. Robinson and Geoffrey Godbey, *Time for Life: The Surprising Ways Americans Use Their Time* (University Park: Pennsylvania State University Press, 1997); Michael D. Shook and Robert L. Shook, *It's About Time!* (New York: Plume, 1992); *Information Please Almanac* (Boston: Houghton Mifflin, various years); Tom Heymann, *On an Average Day* (New York: Fawcett, 1989); many articles in *American Demographics.* Often, a wide range of estimates is available for any one use of time.

So, although I provide a few specific citations, I'm not trying to offer scholarly backing for the data here. The point is that these statistics run from undependable to meaningless.

THE STATISTICIANS CLAIM M. H. Bonnet and D. L. Arand, "We Are Chronically Sleep Deprived," *Sleep* 18, no. 10 (1995): 108–11.

IN TWO HOURS THE BRAIN'S Robert Henderson, "Go It, Tortoise," *New Yorker,* September 28, 1963, p. 39.

EVENTUALLY, IF THE SLEEP DEBT Stanley Coren, *Sleep Thieves* (New York: Free Press, 1996).

DRIVING A CAR National Personal Transportation Survey, Federal Highway Administration, 1977. Survey of 42,000 households.

FREEWAY SYSTEMS ARE DESIGNED Peter Samuel, *Highway Aggravation: The Case for Privatizing the Highways* (Policy Analysis No. 231, June 27, 1995).

ROADWAY CONGESTION INDEX "Urban Roadway Congestion, 1982 to 1994," Texas Transportation Index Research Report 1131-9, p. 11 (http://mobility.tamu.edu/).

Sex and Paperwork

AMERICANS TELL POLLERS Robinson and Godbey, *Time for Life,* Appendix O: Ratings of Detailed Activities on Enjoyment Scale (1985), Diaries, p. 340.

THE BROADEST AND MOST CAREFUL Robert T. Michael, John Gagnon, Edward O. Laumann, and Gina Kolata, *Sex in America: A Definitive Survey* (Waltham, Mass.: Little, Brown, 1994).

SEX MUST BE SLIPPING IN Robinson and Godbey, *Time for Life,* p. 116.

HOURS THAT HAD BEEN SPENT BY ANGLERS "Reports to Congress Under the Paperwork Reduction Act of 1995," September 1997. Office of Management and Budget, Office of Information and Regulatory Affairs.

THE TIME-COMPONENT OF BURDEN "Development of Methodology for Estimating the Taxpayer Paperwork Burden," Final Report to Department of the Treasury, Internal Revenue Service, June 1988. Arthur D. Little Opinion Research Corporation. IRS Contract No. Tir 83-234 Reference 50042.

LOOKING FOR LOST OBJECTS Dan Danbom, "Get in Line," *American Demographics,* May 1990, p. 11.

EIGHTY-SIX MINUTES [ON-LINE] Andrew Kohut, *Technology in the American Household* (Washington, D.C.: Times-Mirror Center for the People and the Press, 1995).

PEOPLE AGED FORTY-FIVE AND UP "SoftUsage Report," Media Metrix, San Jose, Calif., February 1998.

NINE PAINFUL MINUTES A DAY Bob Tedeschi, "Report Puts a Number on the World Wide Wait," *New York Times,* August 8, 1998.

I SAW BILL GATES DEMONSTRATE John Dodge, "A Contrarian View of Windows 98: I Like It," *PC Week,* April 27, 1998, p. 3.

USING SOPHISTICATED TIME MAPPING Ivan Seidenberg, address to Massachusetts Software Council, April 24, 1998 (http://www.ba.com/speeches/1998/Apr/19980427001.html).

Modern Conveniences

YOU NEEDN'T MIND Randall Jarrell, "A Girl in a Library," in *Selected Poems* (New York: American Book–Stratford Press, 1969), p. 16.

FOUR AND A HALF HOURS ON HOUSEWORK M. Marini and B. A. Shelton, "Measuring Household Work: Recent Experience in the United States," *Social Science Research,* 1993, p. 361.

HOUSEWORK NUMBERS John P. Robinson and Melissa Milkie, "Dances with Dustbunnies: Housecleaning in America," *American Demographics,* January 1997, p. 36. Comparisons with and without appliances are corrected by the researchers to adjust for age, marital status, and other factors.

Jog More, Read Less

SIXTEEN MINUTES READING Rebecca Piirto Heath, "In So Many Words: How Technology Reshapes the Reading Habit," *American Demographics,* March 1997.

EXCESS QUANTITIES OF OLD MERCHANDISE Confidential memorandum from a nationwide retail chain, August 1998.

BECAUSE IT HAS MADE INFORMATION Gerry Marzorati, "Quibbles and Corrections," *Slate,* October 19, 1998.

LIFE IS SO SHORT Sarah Kerr, "Mediocrity and Snobbery," September 2, 1998 (http://www.slate.com/Code/BookClub/BookClub.asp).

GREAT MOLOCH, EVIL BAAL Russell Baker, "Busy, Busy, How Romantic," *New York Times,* June 2, 1990.

RICHARD SANDOMIR DESCRIBED Richard Sandomir, "Let's Dawdle, Spit and Play Ball," *New York Times,* May 30, 1995, p. B1.

THE 800-POUND GORILLA Robinson and Godbey, *Time for Life,* p. 20.

Eat and Run

DIRK JOHNSON ASSEMBLED Dirk Johnson, "Do One-Stop Cereals Really Save Time?" *New York Times,* January 17, 1999.

A SEVENTEEN-YEAR-OLD SAYS Molly O'Neill, "Teen-agers Are Reshaping American Eating Habits," *New York Times,* March 14, 1998, p. 1.

PIZZA SERVER http://www.ecst.csuchico.edu/~pizza/intro.html.

How Many Hours Do You Work?

A GENERATION AGO AMERICANS David Nye, chair of the Center for American Studies, Odense University, Denmark, September 1, 1996.

WE HAVE BECOME A HARRIED Juliet B. Schor, *The Overworked American* (New York: Basic Books, 1991), p. 24.

PLAYING SOLITAIRE AND MINESWEEPER Art Isalo, "Improving Billing Productivity" (http://www.mindspring.com/~italco/prdctvty.html).

PROFESSORS AT PENN STATE Robinson and Godbey, *Time for Life,* p. 61.

RAT-RACE EQUILIBRIUM Constance B. DiCesare, "Is the paper chase a rat race?" *Monthly Labor Review* 120 (April 1, 1997): 40.

HOW DOES THE ORGANIZATION Rosabeth Moss Kanter, *Men and Women of the Corporation* (New York: Basic Books, 1977), p. 64.

PACK IN TWO LUNCHES William L. Hamilton, "Power Lunch Is 2 in a Row, Just New York Minutes Each," *New York Times,* November 11, 1997, p. 1.

IT BOOSTS THE CREDIBILITY Michael Lewis, "25-7?" *Forbes ASAP,* November 30, 1998.

DOWN, OVERALL, ABOUT TWO HOURS Robinson and Godbey, *Time for Life,* p. 83.

PEOPLE THINK THEY KNOW Ibid., p. 81.
SO, NOT INCLUDING THAT ONE HOUR Ibid., p. 82.
MY ESTIMATES, WHICH ARE BOTH Schor, *The Overworked American,* p. 177.
THE GROWTH OF WORK Ibid., pp. 13, 12, 7.
WORK ITSELF HAS BEEN ERODING Ibid., p. 161.
MEN ARE ENSNARED Ibid., pp. 148–49.

7:15. Took Shower

SOCIAL MICROSCOPE Robinson and Godbey, *Time for Life,* p. 5.
5:45 AM–6:00 AM Ibid., p. 64.
WHAT ARE YOU DOING IS Ibid., p. 65.
SOMETIMES AMERICAN CULTURE Ibid., p. 37.
ONE OF THEIR OWN COLLEAGUES Ibid., p. 26.

Attention! Multitaskers

IT'S HARD TO GET AROUND Earl Hunt, quoted in Amy Harmon, "Talk, Type, Read E-Mail," *New York Times,* July 23, 1998, p. G1.
I'M STILL LISTENING Mark Prensky, quoted in Hal Lancaster, "Today's Free Agents Work at Twitch Speed: A Buzzword Glossary," *Wall Street Journal,* October 7, 1998, p. B1.
VISUALLY LOW IN DATA McLuhan, *Understanding Media,* p. 313.
TV WILL NOT WORK AS BACKGROUND Ibid., p. 312.
EIGHT MILLION AMERICAN HOUSEHOLDS "Home Tech," report by Media Metrix, San Jose, Calif., September 1997.
A CHILD MIGHT SIT Robinson and Godbey, *Time for Life,* p. 40.

Shot-Shot-Shot-Shot

IN 1982 PAULINE KAEL Pauline Kael, "The Road Warrior," *New Yorker,* September 6, 1982. Reprinted in *For Keeps: Thirty Years at the Movies* (New York: Dutton, 1994), p. 954.

OUR OWN EVER-GROWING *New Yorker,* June 9, 1997, p. 108.
WHO'D HAVE THOUGHT THAT *Washington Post,* January 15, 1999, p. C1.

Prest-o! Change-o!

POINTLESS BUT INTENSE Saul Bellow, "An Unbearable State of Distraction," public address, November 9, 1989 (http://shango.harvard.edu/~ksgpress/ksg_news/transcripts/bellow.htm).
DID NOT TEST, THOUGHT IT *Consumer Reports* 20 (November 1955): 53.
THEY APPROACH THE AIRWAVES Robert Levine, *A Geography of Time: The Temporal Misadventures of a Social Psychologist* (New York: Basic Books, 1997), p. 45.

MTV Zooms By

THE EAR RECEIVES FLEETING Tony Schwartz, *The Responsive Chord* (Garden City, N.Y.: Anchor Press/Doubleday, 1974), p. 12.

Allegro ma Non Troppo

DEATH EVERYWHERE Lewis Thomas, *Late Night Thoughts on Listening to Mahler's Ninth Symphony* (New York: Penguin, 1995).

Can You See It?

MOVIES AT A THEATER TAKE Coupland, *Microserfs,* p. 258.
AN ADAPTATION OF A SEQUEL Woody Allen, "Celebrity," screenplay, 1998.
FAST CUTTING HAS ARRIVED Stephens, *the rise of the image,* p. 148.

REVELATORY, NARRATIVE-CHALLENGING Ibid., p. 82.
LET'S GIVE OURSELVES CREDIT Ibid., p. 154.

High-Pressure Minutes

WHOEVER HAS TO WORK De Grazia, *Of Time, Work and Leisure,* p. 36.
THE IDEA THAT THE POOR Bertrand Russell, "In Praise of Idleness," 1932.
THE CONTROLLERS CURSE Darcy Frey, "Something's Got to Give," *New York Times Magazine,* March 24, 1996, p. 42.

Time and Motion

ELIMINATE UNNECESSARY STEPS Mary Wright and Russel Wright, *Guide to Easier Living* (New York: Simon & Schuster, 1950), p. 125.
WHEN THEY WATCHED THE WORLD Robert Kanigel, *The One Best Way: Frederick Winslow Taylor and the Enigma of Efficiency* (New York: Viking, 1997), p. 313.
IN REACHING THE FINAL Quoted in ibid., pp. 339–40.
TO RECORD THE TIME AND PATH Frank B. Gilbreth, "Motion Study in the Household," *Scientific American,* April 13, 1912, p. 328. Quoted in Stephen Kern, *The Culture of Time and Space, 1880–1918* (Cambridge: Harvard University Press, 1983), p. 216.
OPEN AND CLOSE FILE Levine, *Geography of Time,* p. 71.
EACH DAY WE REAP Kanigel, *The One Best Way,* p. 17.

The Paradox of Efficiency

WE ARE NOW NEAR THE TIME "Coming: The 12-Hour World," *Nation's Business,* August 1969, p. 32.

LOOSELY COUPLED SYSTEMS Charles Perrow, *Normal Accidents: Living with High-Risk Technologies* (New York: Basic Books, 1984), p. 92.

LOST GROUND CAN BE REGAINED President Franklin D. Roosevelt, Message to Congress, January 6, 1942.

NOT ONLY SAVES BUT ALSO GAINS *Cosmopolitan,* May 1942, p. 102.

365 Ways to Save Time

The numbered tips in this chapter can be consulted in Lucy J. Hedrick, *365 Ways to Save Time* (New York: Morrow, 1992).

FILE FOLDERS, TOPIC BY TOPIC Jeffrey J. Mayer, *If You Haven't Got the Time to Do It Right, When Will You Find the Time to Do It Over?* (New York: Simon & Schuster, 1990).

ANOTHER PROPOSES A SYSTEM Alex Mackenzie, *The Time Trap* (New York: Amacom, 1991).

CARVING OUT "HIDDEN TIME" Hedrick, *365 Ways to Save Time,* pp. 155, 326.

The Telephone Lottery

THE AUTHORS OF A SERIES Woody Leonhard, Lee Hudspeth, and T. J. Lee, *Outlook Annoyances* (Sebastopol, Calif.: O'Reilly, 1998), p. 9.

ONCE IT BECOMES ME Scott Klippel, quoted in Stephen Manes, "From Digital Frustration to Small Claims Court," *New York Times,* June 24, 1997, p. C7.

AN ESTIMATED THREE BILLION http://www.badsoftware.com/badindex.htm.

Time Is Not Money

HE THAT CAN EARN Benjamin Franklin, "Advice to a Young Tradesman," 1748.

ALL-NIGHT BANKING Mark O'Donnell, *Getting Over Homer* (New York: Knopf, 1996), p. 54.

IF A DEVICE WOULD SAVE Henry Ford, *My Life and Work* (New York: Arno Press, 1973), p. 77.

SAMPLE MEAN TIME TO FASTEN Clifford Winston and associates, *Blind Intersection? Policy and the Automobile Industry* (Washington, D.C.: Brookings Institution, 1987).

THE IDEA IS THAT A MAN Ford, *My Life and Work,* p. 82.

TIME ISN'T REALLY MONEY George Lakoff and Mark Johnson, *Metaphors We Live By* (Chicago: University of Chicago Press, 1980), p. 13.

FUMIO KOMATSUZAKI Yumiko Ono, "We're Eating Out Tonight, So Please Bring a Stopwatch," *Wall Street Journal,* December 23, 1998, p. 1.

DISNEY WORLD REPORTS June Kronholz, "Families Feel Forced to Live Their Lives Eight Months Early," *Wall Street Journal,* November 20, 1997.

Short-Term Memory

THERE IS NO PRACTICAL OBSTACLE H. G. Wells, "World Brain: The Idea of a Permanent World Encyclopaedia," contribution to *Encyclopédie Française,* August 1937.

The Law of Small Numbers

Some specific number sequences are from the ultimate source on these matters: N. J. A. Sloane, *An On-Line Version of the Encyclopedia of Integer Sequences* (http://www.research.att.com/~njas/sequences/eisonline.html); N. J. A. Sloane and S. Plouffe, *The Encyclopedia of Integer Sequences* (San Diego: Academic Press, 1995).

PSEUDO-EVENT Daniel Boorstin, "A History of the Image: From Pseudo-Event to Virtual Reality," *New Perspectives Quarterly* 11, no. 3 (Summer 1994); Daniel Boorstin, *The Image: A Guide to Pseudo-Event in America* (New York: Atheneum, 1987).

RICHARD K. GUY Richard K. Guy, "Graphs and the Strong Law of Small Numbers," in *Offprints from Graph Theory, Combinatorics, and Applications,* ed. Y. Alavi, G. Chartrand, O. R. Ollerman, and A. J. Schwenk (New York: John Wiley, 1991), p. 597.

1, 1, 2, 5, 14 To express the postage-stamp rule with a bit more rigor: "The number of ways of making n folds in a strip of $n + 1$ postage stamps, where we don't distinguish between the back and front, or the top and bottom, or the left and right of a stamp." Ibid.

THERE AREN'T ENOUGH SMALL Richard K. Guy, "The Strong Law of Small Numbers," *American Mathematical Monthly* 95 (1988): 697–712.

IS REZULIN THE NEW "Avoiding Trademark Trouble at FDA," *Pharmaceutical Executive,* June 1, 1996, p. 80; Anne Federwisch, "Name That Drug," *NurseWeek/HealthWeek,* June 29, 1998.

ALMOST NOBODY WILL BE FAMOUS Woody Allen, "Celebrity."

Bored

PERHAPS YOU CAN JUDGE De Grazia, *Of Time, Work and Leisure,* p. 341.

TO BE BORN IN IGNORANCE Quoted in Patricia Meyer Spacks, *Boredom: The Literary History of a State of Mind* (Chicago: University of Chicago Press, 1995), p. 45.

HUMAN BEINGS NEED NOT LANGUISH Ibid., p. 46.

THE ANKORE OF UGANDA John S. Mbiti, *African Religions and Philosophy* (Garden City, N.Y.: Anchor Books/Doubleday, 1969), p. 25.

ATHENIANS HAD FIFTY Schor, *Overworked Americans,* p. 7.
PRIMITIVES DO LITTLE WORK Ibid., p. 10.
THOSE WHO ARE SEEN SITTING DOWN Mbiti, *African Religions and Philosophy,* p. 24.

The End

IF WE CONTINUE TO FOLLOW Stephen Jay Gould, "Scale Models," *Forbes ASAP,* November 30, 1998.
HUMANS ENDURE A MORE Greg Blonder, "Faded Genes," February 1995 (http://www.genuineideas.com/articles/genes/faded_genes.html).
I'VE ALWAYS MOVED AT A FAST Jay Walljasper, "The Speed Trap," *Utne Reader,* March–April 1997, p. 42.
THE HISTORICAL RECORD SHOWS Stephen Kern, "The Culture of Speed," conference at Netherlands Design Institute, Amsterdam, November 1996.
IF A MAN TRAVELS TO WORK Kern, *Culture of Time and Space,* p. 129.
INSIDE THE PRISON WALLS Maurice Lever, *Sade: A Biography,* trans. Arthur Goldhammer (New York: Farrar, Straus & Giroux, 1993), p. 307.
MONTHS PASSED WITHOUT MY Karel Weiss, *The Prison Experience* (New York: Delacorte, 1976), p. 217.
SOLZHENITSYN, RETURNING D. M. Thomas, *Solzhenitsyn: A Century in His Life* (New York: St. Martin's Press, 1998), p. 136.
PSYCHOLOGISTS HAVE ISOLATED John Cohen, "Psychological Time," *Scientific American,* November 1964, p. 216.
WE EXPERIENCE TIME INTERVALS Gernot M. R. Winkler, "How Many Different Kinds of Time?" lecture notes, October 16, 1986, p. 3.

Index